# FIGHTING FIRE
## IN THE
# SIERRA NATIONAL FOREST

## MARCIA PENNER FREEDMAN

THE
History
PRESS

Published by The History Press
Charleston, SC 29403
www.historypress.net

Cover image by Gay Abarbanell.

First published 2015

Manufactured in the United States

ISBN 978.1.62619.371.0

Library of Congress Control Number: 2014958794

*Notice*: The information in this book is true and complete to the best of our knowledge. It is offered without guarantee on the part of the author or The History Press. The author and The History Press disclaim all liability in connection with the use of this book.

*Fire isn't listening. It doesn't feel our pain. It doesn't care—really, really doesn't care. It understands a language of wind, drought, woods, grass, brush and terrain, and it will ignore anything stated otherwise.*
*—Stephen. J. Pyne*

# CONTENTS

# CONTENTS

# ACKNOWLEDGEMENTS

Writing this book has been an adventure, and I thank those who helped make it so: Richard Bagley of Southern California Edison, who introduced me to prescribed fire and mastication; Burt Stalter of the Sierra National Forest, who drove me through the ravages of the French Fire; Matthew Brown of PG&E, who opened up my world to reforestation of masticated brush fields and of the Bass Lake Dam; the Schroeders and Pickrens, who drove me over their land involved in the Carstens fire; Barbara Thormann, who took me into the lookout and showed me the ropes; Gina Clugston of SierraNewsOnline.com, who introduced me to the French Fire staging area. Without these firsthand experiences, what I imagined could not have equaled what I saw. I also thank Julie Elstner for her keen editing and helpful suggestions; John Mount of Southern California Edison; Erin Potter, Marie Mogge, Mark Smith, Frannie Adams, Paul Waddell, Heather Taylor and Carolyn Ballard of the Sierra National Forest; and Karen Guillemin and Mark Glass of Cal Fire. Lastly, I'd like to thank Anne Lombardo for helping me secure important and useful information.

# LIST OF ILLUSTRATIONS

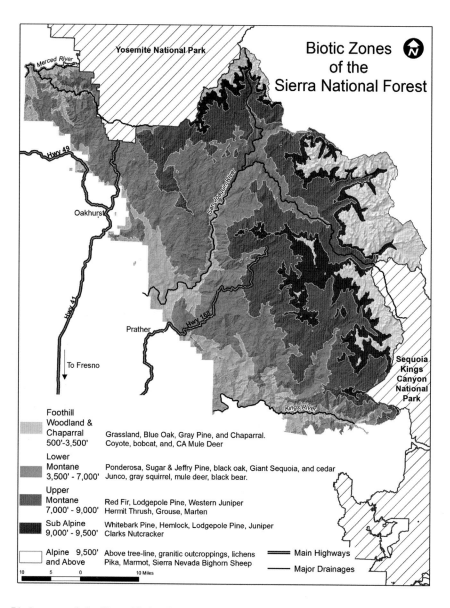

**Biotic Zones of the Sierra National Forest**

Yosemite National Park

Merced River

Hwy 49

San Joaquin River

Oakhurst

Hwy 168

Prather

To Fresno

Kings River

Sequoia
Kings
Canyon
National
Park

Foothill
Woodland &
Chaparral
500'-3,500'
Grassland, Blue Oak, Gray Pine, and Chaparral.
Coyote, bobcat, and, CA Mule Deer

Lower
Montane
3,500' - 7,000'
Ponderosa, Sugar & Jeffry Pine, black oak, Giant Sequoia, and cedar
Junco, gray squirrel, mule deer, black bear.

Upper
Montane
7,000' - 9,000'
Red Fir, Lodgepole Pine, Western Juniper
Hermit Thrush, Grouse, Marten

Sub Alpine
9,000' - 9,500'
Whitebark Pine, Hemlock, Lodgepole Pine, Juniper
Clarks Nutcracker

Alpine  9,500'
and Above
Above tree-line, granitic outcroppings, lichens
Pika, Marmot, Sierra Nevada Bighorn Sheep

Main Highways

Major Drainages

10    5    0          10 Miles

Biotic zones of the Sierra National Forest. *Courtesy of Heather Taylor, former firefighter, now on a Type 1 team as a map person.*

## Part I

# WHAT IS THIS THING CALLED FIRE?

The Sierra National Forest, situated on the western slope of the Sierra Nevada, spans 1.3 million acres—the size of the state of Delaware. It comprises five biotic zones from foothill woodland and chaparral to treeless alpine granite outcroppings. It is a place of extraordinary natural beauty with a terrain ranging in elevation from nine hundred to fourteen thousand feet.

When people choose to live in this area, their choice involves living with wildfire. It's not simply that they learn to clear brush from around their homes or to plan for potential evacuation or that, from time to time, they will have to endure weeks of smoke-filled air because of a wildfire somewhere in the forest. They become acquainted with fire, the living machine that needs oxygen to thrive and maintains a unique and vital relationship to their living world.

## 1

# THE GIVE AND TAKE OF FIRE

T hat fire requires oxygen is something children learn early in life. In a second grade science lesson, a jar is placed over a flaming candle. The flame dies out. The wick smolders and cools. Oxygen, heat and fuel, the fire triangle of the child's lesson, has been fire's story since its appearance on Earth, which, according to fire historian Stephen J. Pyne, was 400 million years ago, when all the ingredients were in place and the triangle completed.

"Of fire's three essential elements," Pyne writes,

> *only the heat of ignition thrived on the early Earth. Oxygen did not begin to collect until the last two billion years, and did not begin to approach modern quantities until roughly 500 million years ago. Land plants suitable to carry combustion did not become abundant until 400 million years ago. Before that time the Earth lacked the means to burn regularly or vigorously.*

For 400 million years, fire has followed a consistent principle: after ignition—which occurs naturally in the form of lightning—sufficient heat, fuel and oxygen must be present if fire is to take hold and thrive. The removal or reduction of any one of these elements during a fire diminishes or extinguishes the fire's energy.

How was it that human beings were able to grasp this principle and make fire on their own? Perhaps they observed the slowing down of a forest fire when the nights turned cool, the halted progress as a fire crept up against a granite wall or the flare up of flames in the presence of a sudden breeze. Whatever it was that allowed them to connect the fire triangle dots, somewhere in the course of their evolution, people learned they could start and stop fires and keep them going. They learned about fuels, figured out which burned hotter, which cooler. They put their fire triangle knowledge to use and manipulated fires of different types—to cook their food, clear their fields, shape their tools and weapons, run their cars and on and on.

Pyne describes the relationship of humans to fire as "species specific." "While all species modify the places in which they live and many can modify fire's environment," he wrote, "only humans can, within limits, start and stop fire at will." Scientists, like anthropologist M. Kat Anderson, propose that control of fire was "the greatest invention in the history of humankind," allowing ancient people to stay warm, cook food and repel predators. Fire might have encouraged them to remain awake after nightfall, contributing to a social life, or prompted them to venture out and settle in otherwise forbiddingly cold areas. "Fire probably had a psychological effect too," Anderson wrote. "Humans knew they had tamed one of the terrifying forces of nature." What an achievement. What a confidence builder. Floods, hurricanes, tornadoes—these couldn't be domesticated. Fire could.

Fire's connection to humans is only one part of its narrative, however. As a force of nature, fire tears through the living world, feeding on biomass and drawing out its energy, in essence, killing it. "Hurricanes, tornadoes, floods, melting glaciers, these are purely physical," wrote Pyne. "Not fires. Fires need life."

Yes, fire consumes life. But in forests where fire is a natural part of the environment, like those in California, fire meets its match. It comes up against life that can resist its onslaught. It encounters plants that even depend on fire to sustain and improve their species. In that sense, fire can be considered neither good nor bad. Rather, as fire ecologist Neil G. Sugihara and his associates explain, in fire-prone plant communities, fire is integral to the natural processes of living matter.

In the Sierra National Forest, many examples exist of plants that have adapted to the presence of fire or use fire for propagation and improvement. The ponderosa pine grows a bark thick enough to protect the tree from fire's destructive heat. The dormant seeds of the buckbrush burst open and germinate with fire, even when the mother plant has been killed. Flowering after fire is enhanced in the mariposa lily and penstemon. In the ubiquitous bear clover, also known as mountain misery, the deep and complicated root system and tenacious underground series of horizontal stems—rhizomes—produce sprouts after the plant's bout with fire.

Devastation of biomass during the French Fire, Sierra National Forest, in the summer of 2014. *Courtesy of Gay Abarbanell.*

Then there is the giant sequoia, the awesome tree that cannot seem to escape numerical descriptors: the tallest (it can reach three hundred feet) and the widest (it is usually fifty to sixty feet around). Considered young even at

During this prescribed burn at Shaver Lake, the fire-adapted trees will not be harmed as a result of the burning. *Courtesy of Michael Esposito, fire ecologist, Southern California Edison Forestry.*

250 years old, many live to be over 3,000 years old. The sequoia reaches full height by 750 years old. The first large limb on the trunk can be as high up as a twelve-story building. In the entire world, they are found exclusively on the western slope of the Sierra.

In the sequoia's relationship to fire, the numbers still boggle the mind. It resists fire by its rutted fibrous bark, which has been known to grow to a thickness of two to three feet. The sequoia defies fire by sending up shoots after fire has passed, the only conifer in the Sierra that sprouts. The sequoia also recruits fire to open its cones. A sequoia cone is the size of a chicken egg—two-and-a-half inches in length. Its seeds have been described as resembling flakes of oat. The cones begin to appear when the tree is fifteen or twenty years old. A mature tree could have 11,000 cones, but some can produce up to 100,000. The cones, each containing an average of two hundred flakes of seed, can remain on the tree for decades waiting for heat from a fire to dry them and open them, releasing hundreds of thousands of seeds in the course of a year. The seeds can travel up to six hundred feet away from the tree as they float to the ground. Once the seeds settle, they need a soft, rich soil in which to embed, a condition brought about by fire that cleans out the pine needles and other debris—duff—around the base of the tree. The seeds will become covered with a tiny layer of the mineral soil, and there, germination begins.

It has been known for a long time that sequoia seeds need fire for germination. A study conducted in 1972 demonstrated the sequoia seed's reliance on fire in a rather interesting way. In that study, three plots of between three and six acres in a sequoia forest were set aside, and prescribed fires of varying intensities were introduced. On a fourth plot, there was no fire. On the fired plots, an average of twenty-two thousand seedlings per acre established themselves, with the highest intensity fires producing the most seedlings. On the plot where no fire was introduced, no germination took place.

If fire can be a friend to the forest, there are instances when a fire can burn so intensely that even the most highly adapted plant is unable to fight back. In such cases, fire can consume most of the organic matter, leaving the plant destroyed and the soil depleted.

Primary succession in action: lichen, moss and grass breaking rock into soil. *Courtesy of Joanne Dean-Freemire.*

At such times, new and hardy life can begin to return. These plants, called "pioneers," improve the poor soil, as if preparing it for the more complex, bio-diverse plants to come. Some of the common pioneer plants in a forest are lichen and moss.

"One of the ways nature has of dealing with bared ground is through what we call secondary succession," explained Joanne Dean-Freemire, retired East Bay Regional Park district naturalist.

*This occurs when soil which was supporting plants becomes bared by serious disturbances, such as severe fire. The "pioneer" plants which recolonize this bare soil may be lichens and mosses but are often fast-growing, aggressive and sometimes non-native seed plants—what we humans may call "weeds." If all goes well, these plants will provide shade for native grasses, shrubs and trees to repopulate over time. If not, they may succeed in crowding out the natives, especially if the soil is continually bared, as with repeated fires.*

John Mount, former forestry manager for Southern California Edison Forestry at Shaver Lake, recalled a time when he came upon an area where a log had burned to ash.

*I was reviewing a burn that was out* [on the Southern California Edison land near Shaver Lake]. *It was early morning. The sun was low. I came across the log, probably an area about two feet by thirty feet, a big old log that had burned up much hotter than the rest of the area. All that was left was ash, nothing there that was organic enough to support life. But the sun was reflecting off of it, and there was a rainbow of colors, and I'm thinking I'd finally lost it. It was absolutely beautiful. So I moved closer and the colors went away because I had moved out of the reflection. But I went and looked closer and closer and there were these microscopic things growing in the ash, like lichens. I didn't have a magnifying glass. I couldn't see close enough to identify them. I was throwing out the thought, "What else is going on in this fire that we have not identified?" That demonstrated to me that the fire burned in different ways so the result of that fire was that different species of wildlife came into that area. I thought, maybe if it hadn't burned on that day in that way, those lichens would become extinct. I don't really know if they were lichens. But it told me we don't know all the little things that are happening out there in the forest, and that you can learn something new every day in forestry.*

## ON ANOTHER NOTE

## Fire Scars: Snapshots in Time

*Although fire did visit almost every landscape in California, it did so with a remarkable variety in frequency, intensity, and effects. California has always been and will continue to be a fire environment unmatched in North America.*
—*James K. Agee*

Someday you might find yourself standing by the stump of a two-hundred-year-old tree. If you're the type who likes to check things out, you'd probably begin counting the rings, the record of its age imprinted on the trunk. And it is likely you'd stop if the distance between the rings narrowed and the counting became tedious or when the rings swirled from a neat path and you lost your way or when they bumped up against a chink or a stain. Leave it to the experts, you'd think, possibly not realizing that there is a field devoted specifically to tree ring counting: dendrochronology.

For the dendrochronologist, a scientist who uses tree rings to date historical and environmental events affecting trees, it is at the chinks and stains and at the squeezed together, sidestepping rings where the fun begins. Like an archaeologist who reconstructs the story of a people from the excavated bits and pieces of their past, the dendrochronologist assembles the life story of trees. The glitches and discolorations, the spacing and swirl patterns of the rings offer clues to the tree's annual ecological and climactic experience. Was it a dry year or a wet year? Was there a period of drought or bug infestation? Did this tree live through fire events, when and of what kind?

Fire scars are the tree's messengers of fires past. They are the healed-over fire injuries etched into the timeline of the tree's rings. They pinpoint the years a tree experienced fire, even hint at the fire's severity. Dendrochronologists map fire scars. With fire mapping comes a backward glance—a snapshot, so to speak—of a single tree's historical bouts with fire or of the patterns of fire episodes, regimes, that visited a patch of trees. Think of mapping a two-thousand-year-old giant sequoia, or a grove of the giants. What a fire story that would tell.

Of late, scientists and foresters have taken a special interest in knowing the stories of the natural fire regimes that existed in the forest before Euro-American settlement at the turn of the twentieth century. That is when the policies of fire suppression were put in place—policies that have persisted for

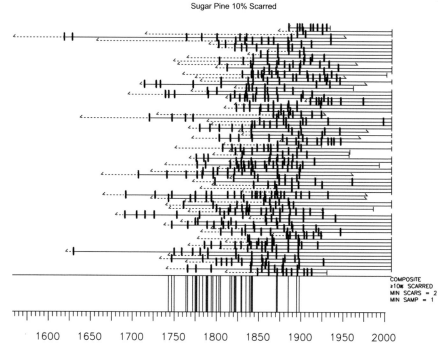

Sugar Pine 10% Scarred

Graph of two-hundred-year history of fire events in sugar pine trees. Fire return is virtually absent after the turn of the twentieth century. *Courtesy of Kevin Krasnow, PhD dissertation.*

a hundred years and have created forests grown thick with understory and choked with tree growth, raising the threat of catastrophic fire.

"They've done this fire mapping all up and down the Sierra," explained Burt Stalter, fuels specialist at the Bass Lake Ranger District of the Sierra National Forest, "and you can see where it stopped, right at the turn of the century when suppression started. You see the fire lines along the chronological line showing a real good pattern of fire, and then, bam, a big open line over a hundred years that shows nothing. No fire history." Not a very good prospect for a fire-dependent landscape.

Fire mapping has been a tool available to foresters for a long time. In 1899, Gifford Pinchot, who at that time held the position of chief forester of the National Forest System, the precursor of the National Forest Service, recognized the significance of fire scars.

*The records of past fires, written in the forest now on the ground, are often decipherable for more than a hundred years back, and in many cases for more than twice that length of time. Such records throw light on the relations of forests and fires as nothing else can, and are consequently the most valuable of all documents upon the somewhat intricate but most important question of the final effect of fire upon the forest.*

But nowadays, the science has advanced. According to Carolyn Ballard, fire management officer at the High Sierra Ranger District of the Sierra National Forest, to achieve a complete portrait of a forest before fire suppression, forest managers need more information than fire maps can provide. They need details about how a forest appeared prior to fire suppression and how fire played on the land.

The situation is complicated by the diversity of the landscape on the Sierra. "The question then becomes, what is the fire regime for a particular piece of land," explained Craig Thomas, executive director of Sierra Forest Legacy. "Is it mostly low and moderate fire effects? Are the fire effects normal for that landscape? For lower elevation forests it's one thing. When you get to the red fir at the higher elevations, with their longer winters and higher moisture, are more intensive fires normal because they're less frequent?" This is the kind of information that forest and fire ecologists are discovering through painstakingly difficult analyses.

Turn-of-the-century open forest. *Courtesy of the Sierra National Forest, Bass Lake Ranger District.*

One goal for forest managers is to apply this information to ecology-based management to restore natural fire regimes in the forest, the mosaics of a vibrant, resilient landscape. "The forests were park-like and open back then," said Richard Bagley, manager of forestry operations for Southern California Edison Forestry. "But there were enough snags and brush, and the ecosystem was healthy and adaptive. The animals had what they needed. They had cover. They had food sources."

The restoration of regimes includes prescribed burning, a carefully planned approach for treating degraded and threatened landscapes. When John Mount went to look over a prescribed burn that Southern California Edison Forestry had done, he saw that the fire they had put down created the desired mosaic on the landscape.

> *It was fantastic. It burned hot in a couple of spots, real hot some places, then other places not at all. Places where the duff had built up really high and you didn't know it, the fire burned up to it, and then there was a one-foot drop down to the mineral soil. You obviously had different things happening on one side of the line than the other. The cyanothous came back, and it was just phenomenal. The result was as it would have been naturally, because it happened during a normal burn period, into the summer, when lightning would have started it and it would have been left on its own to burn.*

# 2

# REGIME CHANGE

## *Practice Makes Perfect—Take One*

*The forests which the first white explorers saw as they landed on this continent and gradually overran it were themselves the successors of others, which, through thousands of years, were burned down at intervals that we can no longer trace. There is but little of all the vast forest area of this country which does not bear, either in actual scars and charcoal or in the manner and composition of its growth, the marks of fire.*
*—Gifford Pinchot, 1899*

What would account for the park-like forests in pre-settlement California? Had they been managed with fire by the indigenous tribes who had inhabited the land until the Euro-Americans came? Or was it solely the result of fire's natural ecology on the landscape?

Some would say the natives did not have the understanding of the complex ecosystems in their environment that management would imply. It has been suggested that the Indians did not have the skill to start and control brush fires on any scale, to know when and how to burn the fire-resistant young plants or handle the conflagrations of the fire-prone older brush. Some believe that Indians did not fire the forest to create healthy resilient regimes but merely to open up paths to make travel and hunting easier or that they hunted and gathered what they needed and moved from place to place when supplies diminished.

It is tempting to accept those points of view. The Indians were a people who made their home in the forest. They did have to clear paths to get around, and they did have to hunt and gather in order to survive. However, many tribes were not nomadic. They established seasonal villages, spending summers high in the mountain meadows and winters in the lower elevations. The local native peoples—such as the Mono, Chukchansi, Choinumni, Miwok and those who have since gone extinct—were the first land stewards of the Sierra forests. Fire was their main tool of choice for vegetation management and improvement. They fashioned their implements and baskets for domestic and ceremonial purposes from the surrounding vegetation. A forest that did not provide the plants they relied on would have left the Indians vulnerable and wanting. So it was incumbent on them to become actively acquainted with how the forest worked and what kept it healthy. In California, that included fire.

In his introduction to Omer C. Stewart's *Forgotten Fires*, anthropologist Henry T. Lewis wrote, "Native Americans would have been seriously endangered had they not understood the effects of and used controlled fires. In terms of what we now know about the ecologies of natural and prescribed fires, the important question is not why hunter-gatherers would have set the fires, but rather why on earth they would not have done so."

Some people would argue that labeling native peoples as "hunter-gatherers" is too simplistic, that it doesn't explain the healthy, park-like condition of pre-settlement forests in California. Until recently, even anthropologists were dismissive of suggestions of indigenous accomplishments with fire. According to Lewis, "The idea that hunting-gathering peoples possessed an elaborate technological knowledge about the use of fire, managed the abundance and sustainability of selected resources, and understood the ecological consequences of what they did was never seriously contemplated until the late 1960s."

But since that time, based on a substantial body of literature—on observations of explorers, miners, ranchers, military personnel and foresters in the mid-nineteenth century, on research and through stories from elders and tribal members—there seems to be a growing consensus that the ecological impact of indigenous fire has been greatly underestimated. Gerald W. Williams, historical analyst with the United States Forest Service, concurred when he wrote, "There is ample evidence that Native Americans greatly changed the character of the landscape with fire and that they had major effects on the abundances of some wildlife species through their hunting."

What are some of the things that have been learned over the past four or five decades about indigenous burning in California? How does this information add to the give-and-take story of fire? In the Sierra forests, the native people adapted to the fire-hungry environment in which they lived. Over the millennia, they learned to embrace fire and to use controlled burns to create and improve the rich habitats that provided their food, medicines and basketry resources. In his essay *Cultural Burning*, Mono (Nium) tribal elder Ron Goode described a characteristic burn plan.

> *Burning an area means burning that particular spot three times in a ten-year period, typically during the first, third or fourth year and again between the sixth and tenth year. An area with severe understory is going to need fire again in the third year. After every fire, grass, seedlings, new shoots and new leaves on burnt bushes all begin to return in the following year. Sometimes the growth is minimal and a passerby would not see the return. For those who live there, the gradual return is obvious. By year two, however, it becomes more evident, and by the third year, one can clearly see the renewed growth. Burning on the third year produces a low intensity fire, and by the fourth year, the brush and or the undergrowth is much thicker and generates a mid-intensity burn.*

In terms of supplying themselves with food, the Indians were agriculturists, not simply gatherers. They studied weather patterns and fuels; cultivated their plants; and tended their vast garden with burning, pruning, sowing, weeding and tilling the soil. They harvested at the right time and in the right amount. In return, the land provided them with leaves such as mint for teas and medicines; grains and seeds for cakes, breads and soups; bulbs; and fruits such as manzanita, elderberry, chokecherry, sourberry, wild strawberry, blackberry, wild grape and gooseberry. The oak tree, which produces acorns, the main staple of the Indian diet, received special fire protection from the native people. By directing the smoke from fires built under the trees, they were able to keep the mistletoe at bay. If you've ever seen a mistletoe-covered oak tree, you have an idea of how thoroughly it can take over a tree.

Even to this day, efforts are made to enhance the output of oak trees with the smoke from prescribed fire. Lois Connor Bohna, a Mono tribal member who gathers and processes acorns and markets them statewide, is always on the lookout for stands of healthy oak trees, free from bugs and mistletoe. "From a good group of oaks," said Lois, "I can harvest up to three thousand pounds of acorns every year."

Tending basketry plants was also an important part of Indian use of fire. The Mono woman, who was the primary basket weaver in the tribe, knew how to gather. She knew when to gather. She understood the elements and the land, such things as that the redbud stick is most pliable and the red color the deepest if cut during the coldest days of wintertime or that sourberry is good when cut in spring as well as winter. Lois is a master basket weaver. She looks for weaving materials that will produce baskets that can withstand day-to-day usage, whether for cooking over hot coals, holding water or carrying a baby. In his book *Fire in the Forest*, Robert Cermak describes his experience talking with weavers. "One learns that they are looking for very specific characteristics when selecting plants for making basketry. Basket maker Norma Turner (Mono), for example, seeks the following qualities when she selects bluebrush branches for the rims of cradle boards, winnowers, and sifting baskets: brownness, roundness, length and no lateral or side branches." These are the characteristics of young, healthy plants that grow after fire has been put on the land and the old, dry, inflexible white sticks have been eliminated.

One might ask how ancient indigenous use of fire in the forest could inform twenty-first-century forest restoration planning. It's been suggested that by discovering the approaches the Indians used in a particular ecosystem, forest managers will be able to tailor their restoration practices to each environment and avoid a one-size-fits-all policy.

"We need to work with very small pieces," explained Carolyn Ballard, fire management officer at the High Sierra Ranger District of the Sierra National Forest.

> *We don't look at a thousand acres and say we're going to do the same thing across those thousand acres. Maybe ten acres here we do something, and ten acres next to it is completely different in terms of how we treat it because it may be habitat or it may be wetter, or it may have burned fifty years ago and it's really dense compared to another part of the thousand acres. So it's looking at it in very fine scale.*

"A major thrust of restoration ecology," wrote M. Kat Anderson in her introduction to Omer C. Stewart's *Forgotten Fire*,

> *is to restore ecosystems to a semblance of the historic structures, composition, and functions prior to major Euro-American settlement and development. Ecological restoration can be defined as the practice of reestablishing the historic plant and animal communities of a given area or region and the renewal of the ecosystem and cultural functions necessary to maintain these communities now and into the future.*

# ON ANOTHER NOTE

## A Chasm of Difference, Firewise

*Our understanding of the historical relationships between fire and society is greatly enhanced if we review the setting in which that society existed.*
—*Scott L. Stephens and Neil G. Sugihara*

Euro-American settlers in the Sierra forest at the turn of the twentieth century encountered fire. Like the native tribes before them, they had to learn to live with the reality of fire, that it would be as much a part of their environment as the snow, the rain, the thunderstorms and the Mono winds. But in contrast to the native Indians, who had thousands of years of forest living behind them and had acquired an understanding of fire's ecological give-and-take nature, the settlers were newcomers, driven by exploration, ambition and a belief in their Manifest Destiny to occupy the land and, by default, its resources.

When they came upon the ponderosas and cedars, the settlers saw lumber. Wildlife carried valuable skins on their bodies. Ore translated into dollars, ounce by ounce. Creeks and waterfalls meant power for turning millwheels and generators. The settlers built towns and businesses, permanent structures that protected them and their provisions from the elements and from predators. They penned their stock behind fences, planted orchards and grew their food on fixed plots of land. Fire became a tool for clearing their fields and something to contend with. After all, one sweep of a wildfire through their land could wipe them out.

For the native Indians, perhaps the skill and knowledge they had gained over the millennia in the use of fire made them feel competent and less vulnerable in the face of a natural wildfire. Or perhaps it was their spiritual connection to nature, and to fire in particular, that allowed them to face the terrors stirred up by a raging mountain electrical storm. Thus, for the Indians, fire was not to be feared. Rather, they viewed fire as their gift to the land for which they received the food, medicines, clothing and other resources that sustained them.

A member of the local Choinumni, a tribe related to the Chuckchansi Yokuts, explained how members of his tribe viewed their connection to the land.

*Sometimes there are areas that have supplied plants to our families for generations. The way we look at it is that the Creator made this certain*

Settlement in Crane Valley, late nineteenth century. *Courtesy of the Sierra National Forest, Bass Lake Ranger District.*

*plant produce so many shoots or acorns for a certain reason. So you have to pay attention to know what the plant needs. The Indian community is like that. They have a lot of respect for Mother Nature, for fire, for the plants that have taken care of their families. For us, Mother Nature is the church. You respect everything about it because it provides for you and keeps you alive. Those who didn't grow up in that kind of environment don't have that spiritual attachment.*

It's not that the settlers avoided fire altogether. On the contrary, in the earliest years of the twentieth century, there were strong advocates among private landowners for the light burning used by the Indians. Henry T. Lewis mentions ten reasons why European explorers and settlers in the United States would resort to burning. Among them were clearing land to initiate agriculture, firing of pastures and rangelands by cattle and sheep ranchers at the onset or end of the growing season and the burning of slash and stumps by loggers to clear detritus and regenerate tree growth. Railroad companies often set fires alongside rail lines in order to reduce fire hazards from passing trains, which, in the early years, lacked spark arresters and tended to cause fires. In the Sierra

"The Speeder" rode rails of the San Joaquin and Eastern Railroad, as mentioned by Gene Rose in his book *Sierra Centennial. Courtesy of the Sierra National Forest, Bass Lake Ranger District.*

National Forest, a special patrol car equipped to ride on rails followed trains to put out any fires they might have started.

But the manner in which the settlers put fire on the land differed from that of the Indians, whose habitat burning was geared largely toward enhancing the quality and quantity of the vegetation. "European settlers used fire indiscriminately to clear areas for farming, ranching, and mining," wrote fire ecologist Jan W. van Wagtendonk. "The impact of such burning was not a concern because vegetation was thought of as a nuisance rather than a resource."

## Part II

# FIRE BEHAVE YOURSELF!

*Fire fighting is perhaps the nearest thing there is to war—and always requires extreme physical exertion, long hours, lack of sleep and constant nervous attention.*
*—Elers Koch (as quoted by Timothy Egan in* The Big Burn*)*

*The development of camaraderie and close personal relationships is a well-known outgrowth of most combat experience. So it was with fire suppression. Men lived and worked together under difficult, often dangerous conditions, but they operated as a team. When they attained their objectives they felt good, they felt close to each other, and thus fire suppression became and remained the most important unifying force for the men and women. On the whole, it was this esprit de corps, this feeling of being the best that might have been the most important factor responsible for fire control accomplishments and the growing strength of fire control in the national forests in California.*
*—Robert W. Cermak*

# 3

# ON THE ROAD TO WAR

*It is common for us to blame our current fire situation on the shortcomings and lack of perspective of past land managers. But that is not the case. The needs and values of society were the driving force of past policies, and those needs and values have changed and will continue to change.*
—*Scott L. Stephens and Neil G. Sugihara*

In 1907, Charles Shinn, the first supervisor of the Sierra National Forest, wrote, "Beyond question the greatest enemy of the forest is—fire." Over the next four years, until his retirement in 1911, Shinn would dedicate himself to building a professional forestry organization, one prepared to do battle with fire. In retrospect, considering the fire-dependent nature of the Sierra forests, one might wonder why such a management decision was made and how such an attitude could have taken hold.

Taking a look back some twenty years before Shinn became supervisor, before there was even a school of forestry in the United States, one can trace a thread of fire protection running through the political wranglings, official documents and personal writings of the players who would eventually establish the National Forest Service in 1905.

Gifford Pinchot, who has been credited with founding the service, would emerge as one of the most active and influential members of this group. In 1890, at age twenty-five, Pinchot returned from a year's study of forestry in France and Germany. Armed with the advice of one of his European mentors to "go home, manage a forest, and make it pay," he was ready to establish himself as a professional forester. But the forestry he believed in, the forestry he had glimpsed in Europe—a timber industry systematically managed for sustainability, productivity in perpetuity and, above all, expanding capital value of the forests—had no practical application in the United States.

The America Pinchot found on his return, as he describes in his memoir *Breaking New Ground*, was "obsessed in a fury of development." It was an America of the individualist spurred on by the nineteenth-century expansionist ethic of Manifest Destiny, which entitled settlers and pioneers to exploit the natural resources of the land. It was an America of giant monopolies that had grown out of the laissez-faire (let it be) economic philosophy of the times, policies that paved the way for industrialists like John D. Rockefeller and Andrew Carnegie and financiers like J.P. Morgan. "The man who could get his hands on the biggest slice of natural resources," wrote Pinchot, "was the best man…Wealth and virtue were supposed to trot in double harness."

In the context of those times, lumbering had grown into one of the largest industries in America, and wood had become a sought-after commodity. Laissez-faire had spread into the hinterlands with investors and speculators in timber engaged in "the most appalling wave of forest destruction in human history," recalled Pinchot.

In the Sierra, for example, in 1880, investors in the Madera Flume and Trading Company zeroed in on the Nelder Grove of the giant sequoias. They purchased twelve thousand acres in the vicinity of the grove, which, according to Pinchot, could be purchased from the government at two dollars and fifty cents an acre under the Timber and Stone Act of 1878. They introduced a steam-driven device to speed up timber production, built a four-mile narrow gauge logging railroad to expand their outreach and, within two years, had logged out the acreage, left the Nelder Grove decimated and abandoned the whole operation.

"This [type of] lamentable and gigantic massacre of trees had reason behind it," Pinchot wrote,

> *Without wood, and plenty of it, the people of the United States could never have reached the pinnacle of comfort, progress, and power they occupied before* [the twentieth] *century began. The American Colossus was fiercely intent on appropriating the riches of the richest of all continents—grasping with both hands, reaping where he had not sown, wasting what he thought would last forever.*

So, under these conditions, could fire protection be viewed as an economic strategy? Had timber become perceived as too valuable, too important to the future of America to let fire have its way with the trees? Had the lumber industry gained enough influence to sway outcomes in Congress?

Yes, yes and yes.

Still, there were other realities that influenced the attitudes and thinking about forest protection. The latter decades of the nineteenth century saw the forests undergo major changes. Those Indians who had managed their forests by fire were gone. Settlers were clearing woods and light burning the newly created fields to start their farms. Miners, stockmen, lumbermen and the railroads were firing the forests for their own purposes. And then, in a bizarre fallout from a new job called firefighting, members of citizen crews were starting fires in order to create a market for their skills.

In essence, mid-nineteenth-century practices with fire, which reflected the independent, Manifest Destiny mindset of the era, were consequential, and they contributed to changing the nature of American woodlands. The park-like open forests of pre-European settlement, forests ecologically prepared to handle fire running through them, had virtually disappeared. Then, on top of it all, with the cut-and-run commercial logging that was taking place, American forests had been transformed into tinder boxes of dying stumps, thick brush and the accumulated slash of twigs and other forest debris, a condition that left them vulnerable to wildfire.

And the wildfires came. Two of the biggest—the Peshtigo, Wisconsin fire of 1871 that killed over one thousand people and burned one million acres and the Hinckley, Minnesota fire of 1894 that killed over five hundred people—remained in the collective memory of the shapers of an American forest plan that seemed to preclude consideration of ecology-based management.

Even the wave of pushback that grew during this time against the abuses of America's natural resources by big business was not strong enough to reverse the momentum toward fire elimination. The environmental organization, the American Forestry Association, for example, which was founded in 1875 for the purpose of restoring healthy forest ecosystems, capitulated in its annual meeting of 1886 when it resolved that "fire is the most destructive enemy of the forest, and that the most stringent regulations should be adopted by the National and State and Territorial governments to prevent its outbreak and spread in timber stands." The preservationist movement, spearheaded by John Muir and his Sierra Club—which advocated for the appreciation and safeguarding of the naturalness, beauty and diversity of nature—had to forge its own path against the conservationist "use" philosophy of Pinchot. Then there were the voices of settlers and lumbermen who advocated for Indian-style, light burning management techniques, a call drowned out by the din of political arguing and policymaking. In sum, it appears that none of the countermovements could stop the creation, in 1905, of the National Forest Service as a fire exclusion organization, with Gifford Pinchot at the helm.

That year, Pinchot produced the United States Department of Agriculture's manual *The Regulations and Instructions on the Use of the National Forest Reserves*, commonly known as the Use Book. The section *Protection Against Fire* opens with the statement, "Probably the greatest single benefit derived by the community and the nation from forest reserves is insurance against the destruction of property, timber resources, and water supply by fire," and further on, "Officers of the Forest Service, especially forest rangers, have no duty more important than protecting the reserves from forest fires." The year following the publication of the Use Book, President Theodore Roosevelt, Pinchot's friend and political ally, was quoted as saying, "The Forest Service has proven that forest fires can be controlled."

It was in this atmosphere that Charles Shinn took over as supervisor of the Sierra National Forest, declaring fire the enemy and developing a management approach that emphasized "the importance of saving every tree, large and small, that we can save from fire, at any cost of time and money."

But Washington, D.C., was far from the Sierra National Forest, where Shinn was trying to figure out how to protect one million acres of timber with untrained firefighters and an inadequate supply of tools. Shinn's wife, Julia, who served as her husband's office clerk, described what she called "negotiating the vagaries of Washington." Despite the Use Book directive that every ranger should "always carry at least shovel and ax during all the dangerous season," for example, Shinn's request for tools was turned down by D.C. because official records showed the Sierra to have sufficient tools. "We laughed," Julia wrote, "and made or bought our own [tools]."

"We put out fires with hoes, rakes, shovels, matches [for backfiring], bits of sticks, or our naked hands," wrote Shinn in 1907, with a hint of admiration for his firefighters. There might even have been a touch of fatherly pride, maybe awe. Shinn was, after all, up in years, a city person, intellectual and scholarly. This contrasted with his corps of firefighters, made up in large part of young Indians and local settlers schooled in the classroom of "hard knocks and experience," as one of them described it. "When a real mountain fire comes," Shinn wrote, these "untaught, unconscious heroes seize it, shake it, choke it into silence and oblivion; then stagger into camp, or drop exhausted on the burnt edges of their fire-line." For Shinn, exposure to this collection of men appears to have been a source of amazement and amusement. "We burn our clothes, singe our beards, blister our faces, ruin a seven dollar pair of boots, lose our voices in the hot smoke, go short on grub, discover a Gargantuan thirst."

On the other side of the firefighting crew, however, were the rangers, the salaried civil servants who were held to a higher standard than their counterparts. In line with Shinn's plan to inject professionalism into the Sierra organization, rangers were expected to learn the forest regulations as outlined in the Use Book and to pass the Civil Service exam initiated by Pinchot.

The story of Roy Boothe, one of the earliest Sierra rangers, provides a snapshot of the professionalism that developed under Shinn. In Boothe's 1940 memoir, he tells of his experience during the winter of 1906, soon after the release of the Use Book. Mariposa district ranger Joe Westfall was enticing him to seek work in the Sierra. "Ranger Westfall was the typical Western cowpuncher type," recalled Boothe, "reckless, self-reliant, and extremely interested in this new job of his." Boothe attended what he described as "informal night school classes" at Westfall's home during the winter months, when the Mariposa District was closed down. The only other person attending the class was a schoolteacher, "who probably realized better than we our shortcomings in Spelling, Mathematics, Spanish, and History, etc." wrote Boothe. But it wouldn't be too much of a stretch to imagine Shinn actually arranging for Westfall to hold these sessions as part of Boothe's recruitment. After the classes, and even with his acknowledged academic deficiencies, Boothe managed to pass the exam and was hired into the service. A year and a half after Boothe attained civil service status, Shinn sent him to represent the Sierra Forest in a ranger school he had organized, one of the first ever to be held. Within a short time, Boothe was promoted to the position of head ranger of the King's River District of the Sierra National Forest, and years later, he would go on to become supervisor of the Inyo National Forest.

Although the ranger's work in these early years encompassed all facets of Sierra management, firefighting composed the most comprehensive and intensive aspect of his job. In those early years, the notion of firefighting as warfare made its way into the National Forest Service. In 1911, the associate forester for California, Coert DuBois, wrote an official United States Department of Agriculture Forest Service document in which he declared, "There is such a close similarity between the task of controlling a forest fire and of checking the advance of a hostile force that military methods can be studied and, in many cases, applied in a more or less modified form." In California, rangers played the "forest-fire game," fashioned after the army officers' war game. Audie Wofford, one of the first rangers to pass the civil service exam and who, within a year, advanced to the rank of chief ranger of the North Fork District, rallied his crew to work with a call to "get 'em early when they're small."

And off to war they went, excited at the "awe-inspiring sight of a fire spreading rapidly through the brush or timber country," wrote Boothe, who noted that fire had a way of arousing men "to do superhuman things, and to have a desire to continue to work for long periods without rest or food, and with even insufficient water. They hated to give up and acknowledge defeat."

In addition to the actual work of suppressing fires, rangers took to climbing to high places to spot for them in the distance. They built fire breaks strategically placed along ridgetops, which had the potential of stopping a future fire's advancement. They cut trails to make the backcountry accessible during wildfires, and they patrolled in search of fires for days and months on horseback, carrying their supplies and equipment with them.

Ranger B.H. Mace, in a 1911 issue of *Sierra Ranger*, the in-house publication of the Sierra National Forest, described the emergency fire packhorse as a strong and good worker, well broken and a good rustler. The saddle would "not be chosen because of its extreme cheapness," he wrote. The ranger would carry water bags, a camp cooking outfit and provisions for two or three days. Tools would be "determined by the locality, with a variety kept ready; a sharp axe and small spray pump or chemical extinguisher."

A large part of the ranger's job, as spelled out in the Use Book, was geared toward educating and enlightening the public about safe and proper use of fire. The ranger was authorized to issue warnings and citations or even eject a person from the forest. The writers of the Use Book cautioned, however, that this part of the job was to be carried out with "utmost tact," cheerfully and politely. Resentment and impatience with the forest administration from settlers and users was to be avoided. A ranger was expected to "handle the public without losing his temper or using improper language."

There is a touch of irony in this directive. Boothe writes about the almost universal apprehension and suspicion of the Forest Service in its early years. Gene Tully, the range manager for the district, described his job as "a hard, lonely and sometimes dangerous life. Sudden illness, a fall, weather, slides and the threatened vengeance of a resentful stockman made the early rangers always watchful."

A pre-Smokey fire-safe message to the public. *Courtesy of the Sierra National Forest, Bass Lake Ranger District.*

Julia Shinn, in a letter written to a friend several years before her death, described the situation when she said, "We were the first to try to stop indiscriminate use of the forests—a use for pasture, for fence-posts, even for lumber, which the settlers had taken for granted were anybody's. The mountain people were simple, friendly folk, urging us to stop for dinner, if we came along near noon, but hotly argumentative on the question of restricted use of the forest area. It took years for them to learn that the national forests belonged to the nation, not just to the neighbors."

But the antagonism on the part of the mountain citizens extended beyond forest use and ownership. The old issue of light burning management lingered. "The 'Indian way' of low intensity burning of meadows and forest understory appealed to many settlers who saw for themselves what happened once those fires ceased," wrote fire historian Stephen Pyne.

*They saw forests thicken into "jungles" with fresh saplings and poles; they saw (they believed) a worsening of pests; they watched fire control become more difficult, no longer a matter of beating out creeping surface fires with burlap sacks or simple backfiring. They pointed out that the magnificent forests for which loggers lusted had survived Indian burning and had perhaps resulted from it. They wanted those practices to continue. They wanted light burning to become the basis for forest protection.*

Not only did the farmers, ranchers and sheepherders call for burning as a means of improving their fields and meadows and of reducing the fuel load in the forests, but Sierra rangers and other prominent people could also be counted among its advocates. Range manager Gene Tully is said to have broached the subject of light burning management on a campout with Gifford Pinchot when he visited the Sierra. He pointed out the benefits of light burning to the superintendent. "Pinchot listened," wrote Gene Rose in his book, *Sierra Centennial*, "but didn't say much. The next day they rode back to North Fork, and Pinchot went on his way."

From a management point of view, light burning stood in direct contrast to the approach being adopted in the Sierra Forest, so it isn't entirely surprising that Pinchot would not want to discuss the issue with Tully. If the organization were to advance, if the community were to become one with the administration, if the rangers were to unite in their support of fire exclusion, it seemed that the issue had to be put to rest.

This wouldn't happen for several decades, however, but the seeds for tamping down the light burning argument were planted in the early years of the Sierra Forest. In fact, the war against this management approach might even have begun years before, unwittingly, with Pinchot himself. In his 1899 article *The Relation of Forests and Forest Fires*, Pinchot's discourse on his brilliant personal observations of forest fire ecology, one finds hints of the rationale for the future fire suppression policy of the National Forest Service: fear of fire and economic considerations. Although Pinchot made the connection between diminished forest health and fire exclusion, he could not justify allowing fire to run.

*In a word, the distribution of the red fir in western Washington, where it is by all odds the most valuable commercial tree, is governed, first of all, so far as we know at present, by fire. Had fires been kept out of these forests in the last thousand years the fir which gives them their distinctive character would not be in existence, but would be replaced in all probability by the hemlock, which fills even the densest of the Puget Sound forests with its innumerable seedlings. I hasten to add that these facts do not imply any desirability in the fires which are now devastating the West.*

In the years following the publication of Pinchot's article, the ecology of fire does not appear in his conservation management approaches. Protecting timber became the major focus. Then, in 1907, Shinn added to the official line with his defamation of what he termed "the Piute system of forestry"—light burning—and his characterization of Indian burning methods as un-American.

"I hate to mention it here," he wrote in 1907,

> *but they were not American home-builders, the planters of gardens, the fencers of fields, the users of wood lots. They were the children of Nimrod* [the great grandson of Noah, originally meaning "a great hunter," later taking on the meaning of "inept person"]; *they desired clear spaces under the great pines in order to aim their obsidian-tipped arrows. For that, and that only, they flung wide their fires in forests in the Sierra.*

## ON ANOTHER NOTE

### *Wind: Fire's Master and Slave*

*Every fire crew boss needs to have a good knowledge of fire behavior if he is to be left on his own responsibility.*
*—A.A. Brown*

In 1923, Roy Boothe was put in charge of a crew of firefighters on a canyon wildfire. For nineteen days, the crews fought the fire, challenged by rough and inaccessible terrain and changeable wind conditions. On one of the nights, "a high wind came up," Boothe wrote in his memoir, "carrying the fire completely out of our control and spreading it down the steep mountainside and canyons." In another entry, he tells of a severe wind that "fanned up some burning embers from some place on the mountain above us, and the fire was off again in a mad race down the mountainside."

Boothe's description of the windblown fires he encountered expresses what firefighters through the ages have experienced: the power of wind to dictate the behavior of a fire. For example, erratic winds are especially common around thunderstorms, generating very strong gusts where it's possible for one end of a fire to have thirty-five-mile-per-hour winds and the other end forty-mile-per-hour winds. When fire begins to act at the whim of the wind, it can become dangerous, mostly because of its unpredictability. "That's the kind of thing that gets our guys killed," said Burt Stalter. "They make a decision to be in one place at a certain time, and then the wind will change unexpectedly, and they get caught."

If it sounds chaotic—and it is, but only because of its randomness and because a fire manager or crew boss cannot plug these winds into a firefighting plan. He or she can only anticipate them and react on the spot.

There are winds, however, that managers can rely on, natural wind patterns that can be drawn on for making firefighting decisions. For example, the normal uphill movement of winds during daytime and the reverse tendency at night provide clues to the direction the fire will most likely travel under normal circumstances. Then there are the prevailing winds, those that tend to blow pretty much from the same general direction within a particular landscape. These also present fire managers with a certain amount of predictability as to a fire's behavior. Winds channeled down a river canyon can be expected to magnify and jet through, as if released from the pinched end of a garden hose. Knowing these things provides a degree of consistency and certainty for the fire manager and crew boss, something on which to grab. "It's nature," said Mark Smith, battalion chief for the Bass Lake Ranger District. "And those things don't change. A wind blowing through a canyon, probably it's blowing there right now."

Sometimes, though, fire seems to take on a life of its own when it is neither blown around by a sudden burst of wind nor moving under the direction of Mother Nature's wind patterns. It's during these times that fire shows what it can do when it asserts itself and unleashes its power. All it needs is the opportunity to get so hot that it starts feeding on oxygen from all sides. As it sucks the air in, an updraft is created from the inward gusts of wind, propelling the flames as through a chimney hundreds of feet into the air. The wind whirls around the burning column, counterclockwise, generating a hurricane-like firestorm with winds that can travel ten times faster than the surrounding winds and with temperatures that can climb to above three thousand degrees Fahrenheit. Embers and burning limbs and branches—firebrands—tossed out the sides of the storm can take the fire in new directions. At such times, no amount of strategic planning will matter. No bulldozer, no airplane, no hotshot crew will slow it down.

During a firestorm, plumes of white pyro cumulus clouds that look like thunderheads—known to billow as high as eight miles into the air—can form as the fire enters the cooler upper atmosphere. There, the moisture in the fire will turn to rain. For those who live on the periphery of wildfire country, who can read the signs of fire—the orange sun, the perpetually smoky air—a thunderhead appearing above the horizon is a pronouncement that fire has taken charge somewhere in the forest.

Is it any wonder, then, that firefighters relate to wildfires as if they were alive? They speak of fires running uphill and creeping along the ground, jumping rivers and spotting across fire lines. Fires can throw embers and firebrands. They can escape, even shape, an entire forest. Wildfires are mean and wily and can show exceptional endurance, fierceness or moxie. Fires are full of surprises, an enemy that must be defeated. They are alive, have moods, feed on oxygen and suck in the air. Fires are driven.

Shuteye Lookout volunteer Barbara Thormann reflected on her experience sighting fires. "It really does seem like it's a living thing you're dealing with," she said. "It's

Illustration of the formation of a thunderhead during wildfire. *Courtesy of Heather Taylor, former firefighter, now on a Type 1 team as a map person.*

almost as though it's trying to spite you. Jump on that thing. Strangle it. It kind of brings out that kind of defense in you. You want to get it."

Burt Stalter has always related to fire as if it were a dragon. "It breathes," is how he explained it. "You watch it pulse, the wind will stop dead for a minute, like the fire is almost taking a breath. Then whoosh, it really starts going."

John Mount summed it up when he said, "Sometimes wildfires act like ladies and gentlemen, sometimes like the axe murderer."

## 4

# LIGHT BURNING TO THE BACK BURNER

The National Forest Service turned five in 1910. In that fledgling year, the agency found itself facing public scrutiny of its capacity to protect Americans and their forests from fire. On August 20, a firestorm of hurricane-intense winds roared through national forests in three western states—Idaho, Montana and Washington—and before unseasonable rains could douse the flames two days later, the fire had killed almost one hundred people, annihilated seven towns and destroyed three million acres of timber. The storm was said to have attacked with more power than an atomic bomb. All in two days! "By early estimates of the rangers, the fire had burned enough wood to satisfy the nation for fifteen years," wrote Timothy Egan in *The Big Burn,* his account of what has been considered the largest wildfire in American history.

Not surprisingly, this incident brought condemnation from those who had opposed the formation of the Forest Service from the very beginning, those who would like to have seen the public lands restored to the private sector. After the 1910 fire, Senator Weldon Heyburn from Idaho, one of the strongest adversaries of the Forest Service, argued if the users of the forest—the loggers, cattlemen, settlers, miners, railroads—had been allowed to cut trees as needed, the fuel would have been gone and the fire would not have occurred. He targeted the rangers, virtually blaming them for the outcome of the fire. His outspoken criticism, in particular his defamation of the rangers, brought a strong public rebuttal from Gifford Pinchot, who reproached Congress for its insufficient funding of the Forest Service and praised the dedication and skill of the rangers, some of whom had "paid the ultimate price" for their public service, he said.

The time seemed right for the Forest Service. The American people were ready to protect their natural resources. Laissez-faire and Manifest Destiny were giving way to a more progressive conservation view of natural resource management. The year following the Big Burn, as the 1910 fire is commonly called, Congress doubled the Forest Service budget, and a bill proposed by Senator Heyburn to privatize all the destroyed forestland was defeated.

But all was not entirely well. Those who would go on to the highest positions in the service would take with them the memory of the Big Burn and its enormous destruction of timber. They would place fire prevention and timber protection on the top of their agendas. Fire suppression policy would be strengthened, with the timber industry emerging as a beneficiary. According to Egan, "Taxpayers would pay for building roads, scouting the big timber, and snuffing the fires, then offer up trees more than two centuries old for a pittance to the [lumber] industry."

In California, after the big burn, the light burning debate gained momentum. Those settlers and ranchers who still valued the Indian style of vegetation management pushed for the right to burn. Others argued for underbrush burning of the forests to reduce fuels and lower fire hazards. The voices in support of light burning came out strong. But with the fire phobia, timber worship and the need to make the forest pay for itself gripping the agency, equally strong counter arguments were put forward, especially from upper-level managers.

In his 1911 article *Fire and the Forest—The Theory of Light-Burning*, district forester for California Frederick Olmsted makes the case for preservation of young trees to boost timber value and dismisses light burning as too expensive and impractical. In reference to what he calls "the old Indian burns," he adopts a belittling tone reminiscent of Shinn's 1907 article. He refers to the natives as savages and redmen who "burned up the forest." He makes the claim that the Indians "reduced over two million acres of valuable timber lands to non-productive wastes of brush." "This is not forestry," Olmsted wrote. "It is simple destruction."

Taking a little "if only" look backward, one might ask if Olmsted's view would have been tempered by G.L. Hoxie's 1910 article in support of light burning if it hadn't come out the month of the Big Burn. Hoxie was a civil engineer and lumberman whose article advocated fighting fire with fire. Would Hoxie's warning that "prevention of fire may be made so complete as to menace the forests with greater danger than they now incur" have influenced Olmsted's thinking? Would Olmsted still have written, "The accumulation of ground litter is not at all serious and the fears of future disastrous fires, as a result of this accumulation, are not well founded." Probably he would have written that because he had his organization behind him "to keep fires out absolutely," in his words.

The light burning issue surfaced in the Sierra National Forest in 1911 in a series of opinion pieces that appeared in the Sierra's in-house publication, *Sierra Ranger*. In these articles, forest officials, rangers and interested others engaged in an exchange of ideas about chaparral management as it pertained to fire suppression and light burning practices.

Chaparral, the low elevation brush fields at the western edge of the forest, presented firefighters with difficult challenges. The landscape was overgrown, and the fires moved quickly through it, especially in the hot, arid summers. Also, the chaparral bordered private lands, which brought a unique set of problems. If fire jumped the boundary of the forest, which was not an uncommon occurrence, firefighters would have to deal with resentful settlers who "in many cases would rather see the fire burn," wrote ranger B.H. Mace. Plus, there would be the challenge of avoiding conflicts with state fire wardens in charge of the fire on those private pieces of land.

Cooperation between state and federal firefighting agencies was just around the corner. In March 1911, Congress had actually passed a piece of legislation called the Weeks Act, which, among other things, provided for interagency firefighting cooperation. But the act would take some time to implement, and until that happened, the Sierra managers were left grappling with the issues at hand.

Regarding the question of light burning, the Sierra rangers who advocated for brush clearance did it largely to support suppression efforts and to keep the costs of suppression down. There was a glaring absence of ecological rationale in these writings and no mention of habitat improvement except in Roy Boothe's piece. While he admitted that light burning of meadows raised water levels and improved grazing, he downplayed its value because of the potential of flooding from the runoff of the excess water. "I think in a good many cases the people haven't given any serious thought to the after effect of light burning," he wrote, "but have been used to the old way of burning over the country every two or three years, and so they just take it for granted." The small benefit gained in grazing, he believed, could not offset the cost of damages from flooding.

Malcolm McLeod, the ranger who invented the McLeod firefighting tool, also got into a cost effective argument but in relation to suppression. When he recommended burning "a margin as early in the season as practicable of 25–50 feet on each side of all roads and trails," he was envisioning a firebreak that would allow one or two firefighters to start a backfire or "guard the fire break, with as much safety as many men could under the present conditions." McLeod admitted that many seedlings would be destroyed, an argument against light burning he must have heard consistently among forest administrators. But he theorized that the sacrifice of the seedlings would be offset by the saving of time and timber.

Delbert Boothe—whether he was related to Roy Boothe is unclear—also wrote about light burning. "If it means to burn over the country in sections or blocks, in the spring or fall, when the fire will not burn too fiercely," he wrote, evoking Shinn's dismissive description of Indian burning, "I do not believe in it." Delbert Booth did support the light burning of roads, trails and fire breaks for the purpose of expediting firefighting and keeping the costs down. "If the service would burn out its main trails to a width of from 40–50 feet," he argued, "in case of a subsequent fire, one man could back-fire with as much advantage as five or six men under ordinary circumstances." Boothe also recommended burning the old slash left after timber sales, presumably to cut down on the amount of accumulated fuels on the ground.

On top of all the opinions about how to manage the chaparral that were expressed in the *Sierra Ranger* articles, there was a debate over the wisdom of keeping these lands within the forest. The chaparral were troublesome. They grew quickly. They burned hot and fast. Those in favor of eliminating the chaparral argued that it would free up firefighters to concentrate on fires in the timber, plus have the added advantage of appeasing the settlers. Delbert Boothe suggested "it would be better to cut out this brush country or part of it at least, and set the forest line back 3 or 4 miles." Mace thought it was "absolutely necessary to include enough of this brush country within the Forest so that we can get in our best work on a fire before it reached the timber."

As it happened, in 1915, the Forest Service withdrew seventy-seven thousand acres of brushland from the Sierra National Forest. In doing so, it not only freed the firefighters to concentrate on fires in the timber but also tempered the contentiousness that persisted between the Sierra forest and the settlers. Returning the chaparral lands to private ownership must have seemed like tacit agreement by the Forest Service to accept ranchers' and farmers' light burning practices.

The year 1915 was a pivotal one in other ways also. By that time, light burning was being used regularly by ranchers for improvement of forage for their livestock. And there were still several supporters of light burning within the ranks of the Forest Service in California. Roy Headley, who was the assistant district forester at the time, approached the subject of light burning from the standpoint of economics. He produced a manual that laid out in minute dollars-and-cents detail the economics of suppressing fire, emphasizing the potential for reduction of firefighting costs if underbrush were eliminated with controlled burning. Therefore, he supported letting low-intensity fires run their course unless they threatened valuable timber or structures.

DuBois, who by that time had advanced to the position of California district forester, endorsed the new economic approach and its light burning principles. According to Jeffrey Prestemon of the forest economics and policy research unit of the Southern Research Station in South Carolina, the economics of fire suppression put forth by Headley has served as a basis for economic theory of wildfire management and still informs current practice.

Headley's manual makes for an enjoyable and eye-opening read for its thoroughness and breadth. In addition to targeting administrators, for example, there are passages directed to the fire crew itself, with specific directions on various firefighting methods, like throwing dirt, using water and a procedure called "feeling for fire." There are also two pages dedicated to explaining how to perform a backfire.

Under "Suppression Financial Policy," it instructs:

> *Forget the conception of fighting fire. Think of it* [only] *as a job of constructing and patrolling control line. The difference is vital. True, the idea of fighting a fire has attractive elements; there is in it the heroic, the fun of a game, the taking of chances and much of the spectacular. It also has elements which make for waste and inefficiency. Fire suppression, regarded as a rush job of constructing a special type of trail, is without glamour; but this makes for construction of held line with the speed and efficiency which would exist on a rush trail job. This is what is desired.*

For the fire boss, or others directly involved with the firefighting crew, the manual spells out everything from the care of the men on the fire line to managing compensation for injuries. There is also a guide for classifying firefighters in terms of their trustworthiness, ranging from "Class A: Men worthy of complete confidence, exerting a well-recognized anti-fire influence, possessing superior ability and power of endurance" all the way down to "Class D: Shirkers; disorganizers; men without proper shoes; men who are from inexperience or inclination disinclined to respect authority and orders; men suspected of incendiary tendencies or of nursing a fire; men not trustworthy for any other reason."

And fire bosses are told under the section "Systematic Resting of Men":

> *In the absence of exact information it may be said that men should be rested five minutes out of every half-hour, and Crew Bosses should be so instructed. "Take a five" is a well-known expression on log drives and other exhausting work and may well be adopted as the regular method of telling a suppression crew to rest until the Crew Boss says, "All right, fellows."*

After that, DuBois decided to go a step further in his support of light burning. He sent Chief Forester Henry Graves in D.C. a proposal suggesting a permit system be put in place that would allow those who own private land within the boundary of the forest to clear brush with controlled burning. David Carle, in his book *Burning Questions: America's Fight with Nature's Fire*, recounts in detail the exchange between DuBois and Graves. DuBois pointed out that the growth of timber and brush resulting from the Forest Service's fire exclusion policy "has choked many old trails and made them well-nigh impassible…It may often happen that young timber and brush may grow up so thickly in close proximity to houses, barns, or other improvements on mountain ranches or mines as to become a serious menace in case of a forest fire."

Graves's response was to tell DuBois to hold back "for two reasons." He wrote:

> *You do not want the public to run away with the idea and the impression to become current—as those opposed to the principle of Government protection would doubtless be pleased to have it become current—that we have executed an about-face and thrown over our old principles. It is easy for us to draw distinctions and see distinctions where the public, or a large part of the public, will not recognize them. Again, you do not want to go faster than you can carry your own organization with you.*

Carle points out that DuBois wrote in the margin of that letter, "a good point." Graves suggested that DuBois go slow, lest they lose "control of the situation." Soon after this incident, World War I broke out. DuBois left his job at the Forest Service and joined the army, and the light burning debate was quieted.

Over the next two decades, fire exclusion policy was to become more and more institutionalized. Legislation was passed that not only strengthened the

Forest Service policy of fire suppression but also enhanced the partnership between federal and state firefighting agencies and between the Forest Service and the timber industry. Those who called for light burning as a forest management approach and not simply for chaparral clearance had lost a great deal of leverage with the transfer of the chaparral lands to private ownership. The chaparral had given them a legitimate and palatable platform for putting forth their arguments for fire on the ground in the forest—reduction of fire hazard, improvement of grazing, public safety, even public relations—arguments that had allowed them to avoid confronting the delicate subject of timber protection directly but which substantiated their position that the forest needed fire.

But it was FDR's New Deal (signed into law in 1933), suggests Stephen Pyne, that solidified the Forest Service's move toward fire suppression. With the coming of the CCC (Civilian Conservation Corps) "suddenly, a civilian army existed to build roads, string phone lines, cut corridors through snag fields, construct enormous fuel-breaks, and fight fire," wrote Pyne. "Almost overnight, an infrastructure for fire protection appeared."

Then, in 1934, when Chief Forester Gus Silcox imposed the 10:00 a.m. policy, fire managers became engaged in an all-out effort against fire. Specifically, all fires were to be controlled by 10:00 a.m. the day after their having been reported. Failing that, 10:00 a.m. would be the goal, day after day, until the fire was contained. This policy would persist for forty years.

CCC crew building a forest road in 1935. *Courtesy of the Sierra National Forest, Bass Lake Ranger District.*

But a mistrust on the part of the public had taken hold at that time which would come back to haunt the Forest Service during the environmental movement of the mid-twentieth century. That was when the public took it upon itself to call on the courts and to lobby Congress to bring about changes in Forest Service policy and practices that it was unable to accomplish through less contentious means.

# ON ANOTHER NOTE

## Lookout Living

*Life on a mountaintop is a little short on the pleasure of Saturday nights. For entertainment there is the pageantry of the weather; wind-swept clouds passing in review and thunderstorms, with streaks of lightning like handwriting in the sky*
—*Jeanne Kellar Beaty,* Lookout Wife, *courtesy of Jim Bates*

"Get 'em early when they're small," was Wofford's command. The Sierra ranger, in 1907, had his marching orders. Get to the fires and put them out before they spread, "no matter what other job you're involved in." But how much territory could one patrolman cover on horseback? How fast could he travel? What if he needed help? And what happens when his other jobs interfere with his firefighting mandate?

Enter the lookout, a ranger perched high above the forest whose job was to spot fires and alert the whole corps. Charles Shinn had a vision for how lookouts would fit into his firefighting plan. "We can hold a rock-shelter near the very top of a mountain peak," he wrote in 1907, "can pack in a ranger with supplies, and let him hold such a lonely fort from July to October." There he would have telephones and maps and other devices for locating fires with certainty, Shinn fantasized. "Perhaps we can give him a heliograph, and let him flash the story to headquarters," bouncing his sunlit Morse code messages around the mountains. "It will be a lonely and wonderful life away up above the clouds, and near to the stars." But in 1907, there was barely a navigable trail to the high country, much less a rock-shelter. No telephones, no locating devices. And heliographs were of limited use, requiring sunshine and clear air to be effective.

Despite the difficulties, or perhaps because of them, spotting fires seems to have sparked the imagination of some of the rangers and spurred them into

Rangers stand beside Shuteye Lookout. Had they been there to board it up for the winter? *Courtesy of the Sierra National Forest, Bass Lake Ranger District.*

action. Some became proactive and sought vantage points on high rocky places to search for fire. Others patrolled for weeks at a time in search of fire.

Gene Rose, in his book, *Sierra Centennial*, tells the story of Ranger Audie Wofford and Fire Guard J. Clyde Werly, who, in 1907, climbed to the top of Shuteye Mountain—altitude 8,358 feet—and took up vigil at the summit. There, day after day, from early morning to late in the evening, they took turns watching for fire and building a crude trail. That summer, after four fires had been picked up and Shuteye was deemed beneficial, the Forest Service improved the trail to the top of the mountain. Wofford and Werly began a two-year project of building a crude wooden shelter there. Shuteye was the first lookout constructed in the Sierra.

Other lookouts were built, as were guard and ranger stations and forty-eight tool caches equipped with toolboxes. The Sierra entered into a cooperative firefighting agreement with the Fresno Flume and Lumber Company, the Madera Sugar Pine Lumber Company and the San Joaquin Light and Power Company, which established their own firefighting crews prepared to assist the Sierra when needed. Also, by 1912, telephones became part of the lookout communication system, sending the heliograph into retirement.

To this day, after having undergone several renovations, Shuteye serves as one of a triangle of Bass Lake District lookouts with Miami Mountain and Signal Peak, scanning an area that includes the San Joaquin River Canyon,

Shuteye Lookout, 1952. *Courtesy of the Sierra National Forest, Bass Lake Ranger District.*

Bass Lake, parts of Yosemite National Park, the mountain communities of Madera and Mariposa Counties and the Wawona Basin. "Our lookout system is unique in the sense of communication," said Francis Adams, fire prevention officer on the Bass Lake District. "I've had people tell me that when they go to work in one of the other forests in California, you get in your truck in the morning and you don't talk to anybody. At the end of the day you check out. Here on the Sierra, we're a team. We keep in contact. We know where everyone is and what's going on. In other forests people are on their own, and I think, wow, we really are fortunate in our teamwork. I get to know the people and recognize their voices on the radio. It's fun."

On Shuteye, the lookouts spend a week or two living on the mountain. It's not everyone's cup of tea, to hole up for a week or two in a fourteen by fourteen steel enclosure perched eight thousand feet above the surrounding countryside, scanning the horizon for fire. For Barbara Thormann, who has served as a fire lookout on the Sierra National Forest since 2001, it's something she looks forward to every year.

Following are comments from Thormann about her experiences. In addition to Thormann's impressions, included are comments written by Jeanne Kellar Beaty in her 1953 book *Lookout Wife*, a memoir of the summer she and her husband, Chip, as newlyweds, served as fire lookouts on the Salmon National Forest in Idaho.

Thormann on orienting to the vista:

*Basically, the fire lookout has one job, all day, every day: to spot "smokes" and report their location to dispatch. Locating the fire is really not that easy. Even though I had lived in the mountains I was looking at, had backpacked and hiked and ridden horses over every inch of the territory, looking down at them from the top was a different thing. You have your maps, your longitudes and latitudes, and, of course, you have the Osborne fire finder that helps you gauge the distances, but it still fools you. It's a bizarre feeling because you can't really tell. Well, gee, is that two miles or two blocks? You just have to have been there awhile. You gauge it from what you know.*

*In the forest, there are lots and lots of roads you can see on the map. You have to know them. You can send a fire engine to a smoke and if you miss it and they're off on the wrong road, they can't get to it. It's not like, if you send them on the wrong road they can just scoot around on another road. If you're off two miles, that could be serious. No intersections to turn on.*

Beaty on orienting to the vista:

*I didn't think we would ever be able to identify by name all the country around us. But I was wrong. The forest that was once just an illegible sweep of magnificent scenery will never be just scenery again. It was like meeting a beautiful woman. At first you know only that she's beautiful. Then as you come to know her, you see, not the features that first attracted you, but the personality that shines out from her face. And then, if someone should ask you if she were beautiful, you'd say, "Why, I don't know. She's Mary." There is something sad about such familiarity, the blunting of the sensations. I will never again be able to look at that country and think of it, truly, as wild and unknown. I will never be able to live again through the excitement, the fear, the expectation of my first thunderstorm on a mountaintop. Never again will I feel the suspense, the bewilderment of that first fire.*

Thormann on hosting visitors:

*In the past, people would come up and bring friends, and they'd want me to show them the landmarks and you'd get used to seeing them. But that was pretty much it. I've been doing this for twelve years, and I'd say the last two years the number of visitors has really increased. I had 58 people on one three-day weekend. I was astounded.*

Shuteye Lookout with guests, 1913. *Courtesy of the Sierra National Forest, Bass Lake Ranger District.*

*I've had trekkers from all over the world. They come up and love to sign the log. We have logs up there from forever. There are people who come up and want to see the logs from the '50s, the '60s. They say their parents came up. So it's kind of fun.*

Beaty on hosting people:

*The Handbook for Lookouts and Firemen, a copy of which reposed on every fire finder, had explicit instructions about eliciting information from visitors. Greet them by announcing your name, it said, and they will probably respond with theirs. It didn't say what to do if the visitor said merely "Ugh!" Extend your hand, tell them how glad you are to see them, the handbook continued, and show them around. It went on to caution the lookout about finding a subject of conversation on which he and his visitor agreed—there must be no arguments with visitors.*

Thormann on nighttime and lightning storms on the mountain:

*Nighttime is wonderful, so beautiful. If it's a full moon you might as well plan to read because you can't sleep. It's funny how I love it now, but the very first night of the very first week of my experience, I went off service with dispatch, and it was time to get ready for bed. And here I am in a glass house. I'm standing there thinking, am I just supposed to undress? The whole world can see. It was the weirdest feeling and I looked around and thought, "Oh, well, if anyone wants to look, fine." There's no privacy. But it's just gorgeous at night.*

*I think the only time it's questionable is when you get thunder and lightning. Oh, man. You hear the boom of the thunder, but the lightning you hear the crackle. It's a crackling noise like static. It almost makes your hair stand on end. It's really loud, and you can't go to sleep. I mean I'm in a metal can. And it's so bright, almost like daylight.*

*When there's lightning, I'm up. I don't have to worry about moving about. I'm grounded. Every one of the lookouts is grounded. We have all wooden floors. The beds are made of wood. Everything. The furniture stands on glass coasters. But I'm up, and I don't hold a radio in my hand. The minute I see a strike, I line it up and I write down what the degrees on the fire finder are so that in the morning I can go back to those spots and see if anything is burning. Lightning is important always. But thunder and wind, I just pull a pillow over my head and say to myself, this lookout has been here since you were born. It's going to be here tomorrow morning.*

Beaty on nighttime and lightning storms in the mountain:

*Sometimes at night, we would sit in the darkened lookout and watch the drama of a gigantic storm cloud far off over another forest. Sudden flashes of lightning would light up first one end of the cloud and then another as the spotlight moved from player to player, revealing the depth of the storm.*

*One night a powerful bolt dropped right across the river about ten miles away and burned in the air for nearly a second. Hurriedly we closed the door and carefully shut all the windows, per instruction on a placard tacked to the side of the firefinder. The card warned us to stay away from all metal objects during the storm and to stand on our small black rubber mat if we had to make any telephone calls. Our telephone did not have a cut-off switch, and we had an open wire leading right into the lookout. Every lightning strike now was marked by jerky rings on the telephone. Back to back we stood on the black rubber pad, our spinal columns shivering at each other, while we tried to keep track of the strikes. They bore down on our tiny lookout, the abrupt bolts of lightning moving closer and closer. I tingled with fear and excitement, wondering how soon the building would rock under a direct hit…*

*There was no time to be frightened. We were as busy as an octopus with a seven-year itch, attempting to get a reading on all the flashes, to stop all the water, to rescue sodden food and clothing and to be on the mat at any time lightning chose to strike the lookout. It was well we were busy or we would have frozen.*

# 5

# FIRE BACK FROM THE FRINGE

*Perhaps the most remarkable of the regulative effects of forest fires relates to the composition of the forest—the kinds of trees of which it is composed and the proportion of each. This effect depends upon the action of fire in combination with the various qualities of resistance which trees possess...*
*The study of forest fires as modifiers of the composition and mode of life of the forest is as yet in its earliest stages. Remarkably little attention, in view of the importance of the subject, has hitherto been accorded to it...*
*For we must clearly realize, before the present subject can fall into its proper sequence, that we have not stated everything when we say that "a given forest is destroyed by fire."*
—*Gifford Pinchot, 1899*

Gifford Pinchot was a natural scientist and a researcher. Through his personal observations, he intuited the relationship between fire and the forest. He believed that research on that topic would be "one of the most fruitful and fascinating of all the fascinating and fruitful branches of forestry in the United States" and that it would "remain to attract and reward the student of this branch." Above all, he valued research as a starting place for making policy.

But Pinchot's life work in conservation and in the establishment of the National Forest Service drew him into politics and bureaucracy. As he had predicted, others pursued forest and fire research. The findings of these scientists would ultimately bring about some of the major changes in forest management policy during the twentieth century, in particular as regards prescribed fire, also known as controlled burning or light burning.

One of the pioneer researchers was Yale professor H.H. Chapman, who Robert Komarek called "the father of controlled burning for silvicultural purposes"—referring to managing fire for tree growth and health and not simply range improvement or fuel reduction. In his 1926 article in the *Yale Bulletin*, Chapman detailed the critical need of fire for regeneration of the longleaf pine of the forests in the southeastern part of the United States. Over a period of twenty years, "Chappy," as his friends called him, would publish numerous articles on controlled burning, making recommendations of fire use for such things as eliminating competitors, controlling brown spot disease and reducing hazardous accumulations of fuels, as well as for promoting regeneration of the longleaf pine.

During those two decades, other researchers studied controlled burning in the Southeast, some conducted in partnership with the Forest Service. In addition to findings on the benefits of fire for longleaf pine regeneration, researchers were demonstrating enhancements in forage for cattle and wildlife and in soil fertility.

In 1923, the Forest Service collaborated with the Bureaus of Animal and Plant Industry in a study on fire effects on forest range grazing and longleaf pine reproduction. After six years of data collection, the research team reported positive results for both grazing and for longleaf pine regeneration. In 1939, six years after receipt of the completed manuscript, the Forest Service published the article *Effects of Fire and Cattle Grazing on Longleaf Pine Lands, as Studied at McNeill Mississippi*. This was the Forest Service's first official recognition of the merits of controlled burning.

There was a parallel story going on during those twenty years of groundbreaking research that would add strength to the controlled burning arguments. As the light burning advocates had predicted, fire-deprived national forests had filled up with underbrush and small trees. They were burning at ever-increasing rates and intensity, with fires that were becoming harder and more expensive to contain. Faced with this reality and with research evidence on the benefits of fire use, forest managers probably had no choice but to pay attention. Some might have even acknowledged that, perhaps, complete fire elimination had not been the wisest forest management choice.

So, in 1943, when Lyle Watts, chief of the Forest Service, was invited by southern foresters to personally inspect the devastation caused by wildfires in the region's national forests, he was receptive. He saw the destruction, realized the loss and acknowledged the need for change. "I must admit that control burning has me somewhat confused," Watts wrote to the southern district supervisor. "However, the way that the big fire substantiated your own judgment of things to happen, within a week after you explained it to me, lends a lot of emphasis to your own ideas." Watts approved controlled use of fire on national longleaf and slash pine forestlands to "get at answers which the land administrators must have."

Meanwhile, in California in the 1940s, except for piling and burning logging slash, controlled burning was not practiced in the national forests, and the technique had little support among the forest officials. As Harold Biswell, one of the pioneers in controlled burning, wrote in his book *Prescribed Burning in California Wildlands Vegetation Management*, "Much confusion and controversy existed [in the early '40s] because the need for fire in forest management and for expertise in burning was little understood. Emphasis had been on fire prevention and suppression. Many foresters frowned on the use of fire, finding it difficult to understand the difference between a wildfire and a prescribed fire."

Perhaps California forest officials were confused and didn't understand, as Biswell allowed. Perhaps other things played into their attitudes. Was it, as Chapman said in 1946, that the Forest Service position reflected "a fundamental lack of trust in the innate intelligence of farm and forest owners?" Or were there other factors that contributed to the forest managers' lack of interest in the subject—like the existence of an entire infrastructure devoted to protecting timber? Consideration of timber preservation and the entrenched beliefs about how to save trees sat at the very heart of the organization and probably deafened forest leaders to new ideas. Also, it is possible that remnants of the old anti-Piute burning arguments that had originated within their ranks continued to hold sway. And there were Show and Kotok's conclusions of fire being a destroyer and a menace. These were respected researchers. Both would move on to high-ranking positions in the Forest Service. They made powerful claims, ones that would be hard to ignore.

Whatever combination of circumstances kept forest officials in California from joining the movement toward fire management change, the result was that research and experimentation with controlled burning in California was not only discouraged but was considered controversial. "It is inconceivable that such important investigations be neglected so long," wrote Biswell. "Most investigators wanted to stay away from this activity because it was too controversial. However, I have always reasoned that if there is controversy about something, it indicates a need for investigation and research."

Biswell brought this attitude to California in 1947 when he left his position as head of range research at the National Forest Service Southeastern Forest Experimental Station in North Carolina and joined the forestry faculty at the University of California–Berkeley. Over his thirty-year career, he faced continued controversy regarding his research on prescribed fire in the forest, his public workshops and field demonstrations. The dean of the School of Forestry admonished him to go slow and be more conservative in his presentations and publications, recommending that he focus on the use of fire in the improvement of range rather than in safeguarding the forest.

"Very soon," he wrote, "I found myself involved in research on the use of fire in Sierran foothill woodland-savanna to improve ranges for livestock grazing and wildlife. I spent many weekends in the foothills working with ranchers in control burning to reduce and manipulate brush."

It was regular practice for private ranchers to pool resources and burn together. "More often than not," wrote Biswell, "50 to 60 ranchers came to a control burn, which might cover portions of several ranches—up to 2,000 or 3,000 acres."

In 1967, Bayard Stone, a longtime resident of Bass Lake and former summer ranger for the Sierra National Forest, participated in what might have been one of the last of those burns. "They were community affairs," explained Stone.

*And the purpose was to create more and better grazing land by burning off scrub brush and allowing the high quality vegetation to grow. As was customary at that time, it took place Labor Day weekend. In preparation for the burn, after gaining a permit to do the burn and guaranteeing that the burn would not escape, the owner of the property spent several days constructing a bulldozer fire line around the two thousand acres to be torched. Ranchers and others living in the community all pitched in to help with the burn. Before dawn on the day of the burn, the men involved gathered at the top end or high point of the area to be burned. A fire boss, not the property owner, was assigned. His first job was to designate two "captains" who were in command of the two sides of the fire. They were to patrol the fire line in jeeps, giving out orders. I was a burner. I had a torch and when the signal was given I began torching along the fire line. As I moved along spreading fire, others would be stationed by the captain along the line to make sure the fire was contained.*

*Meanwhile, the women were back at the ranch house making sandwiches which were given to couriers to deliver to the volunteers along the fire line.*

*Of course, the couriers were men, as at that time women were not allowed anywhere near the fire lines. Later in the afternoon, as things kind of settled down, the men were brought to the ranch house in shifts for a big meal.*

*During the day, there were a few mini emergencies when the fire jumped the fire lines. "Fire in the hole!" would be the shout that went up. As night approached, men would be stationed along the fire line in pairs to make sure the fire was contained. That's when the drinking started. We drove along in a jeep delivering six-packs of beer for everyone. But many of these fireguards had the foresight to also bring along a fifth of whiskey to help quench their thirst after such an arduous day. Needless to say, we were offered many a snort along the way.*

*These community affairs amongst the cattlemen of that era are now a thing of the past. In a sense, they were "environmentalists" as they knew the value of fire to help create better rangeland. They considered themselves cowboys and cattlemen and would probably take issue being referred to as "environmentalists."*

One can imagine why these men would have taken issue with being called environmentalists. They were part of a long history of ranching in the Sierra foothills. It was 1967. They were living through a time of environmental protest and reform not unlike the kind their ancestors experienced in the settler days. These men no doubt had heard many tales of when their families had lost open access to the forest and meadows because of federal regulation.

By 1967, the government had already enacted three laws that encroached on the freedom of these ranchers to light burn their fields: the Air Pollution Control Act (1955), the Clean Air Act (1963) and the Air Quality Act (1967). And although the Environmental Protection Agency (EPA) would not be created until three years later, it is likely these men foresaw the coming of the federal regulatory agency that would further hamper their burning activities. It is also likely that they were able to envision the ultimate demise of their light burning practices.

Ironically, however, they were environmentalists, as were their ancestors. Their fathers and grandfathers had adopted the burning techniques of the Indians and grazed their cattle and sheep on the rich forage of the fields they burned year after year. Their families might even have participated in the original light burning activism that possibly helped bring about the return of the seventy-seven thousand acres of chaparral lands from the Sierra National Forest to the private sector.

These latter-day cowboys and their families had practiced ecology with fire long before the terms "fire" and "ecology" were paired, which occurred for the first time in public discourse in 1962 at a groundbreaking conference at the Tall Timbers research facility in Tallahassee, Florida. At the Tall Timbers gathering, nongovernmental conservationists and resource managers learned about the art and science of controlled burning. Stephen Pyne, in *Tending Fire*, refers to this meeting as "a major forum for an alternative vision of fire"—that is, through the Tall Timber Conference, individuals concerned with resource management would have the opportunity to consider the inclusion of fire ecology into their thinking and planning. This opened up new possibilities for scientific research and for managing forests and fields on private lands. There would be fourteen additional Tall Timber Conferences over the years where people from all over the world exchanged ideas about fire ecology. Biswell wrote in 1989 that the published proceedings of the Tall Timber Conferences were the best source of information on the role of fire in wildlands.

In 1962, public lands began to move away from fire elimination as their sole management approach. In that year, Secretary of the Interior Stewart Udall appointed a Special Advisory Board on Wildlife Management to study overgrazing of elk and other wildlife issues on Yellowstone National Park. He selected Starker Leopold, son of environmentalist Aldo Leopold, to chair the committee. The resulting report, *Wildlife Management in the National Parks*—which has come to be known as the Leopold Report—went far beyond the original intent of the commission and took an ecological stance. The report included a recommendation for a change of focus from fire protection to habitat preservation. In other words, instead of automatic suppression of all fires, controlled burns could be used to shape habitat vegetation in a natural way, calling for "ecologic sensitivity."

The National Park Service adopted the management concept in 1974. The National Forest Service began looking at its fire exclusion program in the early 1970s and by 1978 had abandoned its 10:00 a.m. policy and adopted the recommendation of the Leopold Report. That's when the transition in the national forests began, when a shift from viewing fire as something to be controlled to viewing fire as an ecological component of forest management occurred.

It would probably come as no surprise to learn that the move from fire elimination to inclusion of fire in forest management had met many stumbling blocks in 1978. At that time, the basic understanding and acceptance of fire ecology and its relationship to forest health were just beginning to enter the consciousness of Forest Service officials. Some of the old guard managers were resistant to the new fire management approach. Some were fearful. As Forest Service scientist Robert Mutch explained, it was "easier going to bed at night knowing that you've worked all day trying to put a fire out than to go to bed knowing that you plan to do nothing about it." For those managers who had embraced the notion that all fires were bad, the prospect of lighting up the forest must have been rather worrisome.

On an institutional level, the Forest Service was challenged with revisiting its seventy-year-old fire exclusion policy, which had been driven by such documents as Pinchot's Use Book, DuBois' time-focused firefighting manual, and Headley's economic guidebook—all three written from the assumption that fire would be suppressed. The whole infrastructure of the organization needed to be reshaped to accommodate prescribed fire within an ecologically driven management plan. Above all the issues involved in the transformation, the technique of executing ecologically sound controlled burning proved to be a major challenge for the organization in those early days.

"The real difficulty lay not only in conception, but in practice," wrote Pyne.

> *For it proved surprisingly tough to burn instead of extinguish. The reasons were many, though the essence may boil down to two. First, the fire establishment's infrastructure existed to fight fires, not light them. Society, too, sought fire protection, and considered prescribed burning a token gesture in select sites. The second reason is that no one really knew what prescriptions were appropriate or how, other than scattering fire over a setting like the ashes of a burnt offering, a restored landscape might happen.*

Harold Biswell, three years before his death in 1992, wrote "Prescribed Burning in California Wildlands Vegetation Management" in which he presented an overview that condensed his fifty-year adventure with prescribed burning into seven easy-to-understand chapters ranging in topics from the fundamentals of fire behavior and fires set by Indians to effects of prescribed fire on forest resources. In his chapter on fire management and techniques, Biswell offers concise, step-by-step explanations of the approach to prescribed burning management, from selection and training of personnel to mop-up and monitoring. In the chapter on the positive effects of prescribed burning, forest managers are presented with concrete scientific evidence about the ecological benefits of prescribed burning to soil, water, wildlife, wilderness, timber, forage, clear air and cultural and visual qualities. In his final chapter, "Why Not More Prescribed Burning," Biswell refutes thirteen arguments that would contribute to management's holding back on using the technique.

It's almost as if he was saying to fire managers, ecologists, conservationists, on-the-ground crews and everyone else involved with and concerned for American forests, "OK, this is what I know. Now you take it from here."

## ON ANOTHER NOTE

### Practice Makes Perfect—Take Two

*Once I was accused of making prescribed burning look too easy. As a matter of fact, it is not difficult. But it does require knowledge of basic fire ecology, careful planning, patience, experience, and know-how. It can be hard work, and there may be cases when surrounding fuels and changes in weather make it nerve-wracking.*
*—Harold Biswell*

Just a short thirty-five years ago, when the Forest Service made its turnaround regarding fire, there were still naysayers and doubters about fire's ecological importance in the forest. Today, there is no argument. All agree that fire is a natural part of forests and contributes to their health and resiliency. Fire managers are finding ways to return fire to the forest, one approach being prescribed burning.

As its name implies, a prescriptive burn targets an area that has been diagnosed as needing restoration. Prescribed burning is applied for a variety of managerial purposes—improving wildlife habitat, for one, or managing risk of wildfire by reducing fuels. Prescribed fire can be helpful in restructuring landscapes to establish desired fire regimes where fire will burn through naturally. Overall, forest resiliency to fire, disease, drought and pests is a major goal of prescribed burning.

Burt Stalter, fuels specialist at the Bass Lake Ranger District, incorporates prescribed burning in his fuels management planning but emphasizes that prescribed burning is not meant to replace fire suppression. "There will always be a need for fire suppression," he said.

*But there are things we can do beforehand that can affect the level of suppression needed. Fuels management, in this case prescribed fire, is one*

John Mount strip firing during a prescribed burn. *Courtesy of Southern California Edison Forestry.*

A candidate for prescribed burn. Notice the large amount of slash in the foreground and the thick understory of small cedars. *Courtesy of Burt Stalter.*

The Sierra during prescribed burn. Notice the understory and slash being burned. *Courtesy of Burt Stalter.*

The Sierra after a prescribed burn. Notice the clean forest, the light coming through and the large trees unharmed. *Courtesy of Burt Stalter.*

"tool" that is used pre-suppression. That means we try to treat the fuels before the fires happen. By reducing the amount of fuel—vegetation—in an area or changing the type of fuel in an area, the type of fire behavior can be modified or changed when a wildfire occurs. In areas where the fuels have been modified or changed, fire outcomes can change from negative to positive as well. It can provide firefighters with the ability to gain a handle on a fire with less effort and more success than in untreated or managed areas.

But we can't go out and just burn what we want. There are a lot of things that need to be considered before applying prescribed fire on the ground. We work with an interdisciplinary team of resource specialists to plan out where our prescribed fires are the most needed, where they can be the most effective and can have the greatest benefit. These resource specialists include a biologist, hydrologist, botanist, archaeologist, geologist and silviculturists.

The silviculturists are an integral part of this team. They are the ones who project into the future, telling us what is needed to sustain a healthy forested ecosystem. They inform us where prescribed fires can be conducted without adversely affecting the growth cycle of the trees. They alert us to possible tree damage where young age or small size might leave the tree unable to withstand the effects of the fire. Silviculturists can also tell us when a pre-treatment is needed prior to a prescribed burn in order to reduce negative fire effects.

The kind of burning we do in the Forest Service is aimed at improving the forest. We tailor the burn based on what is needed on the landscape. The goal is to get the fuel conditions to a level where, if a wildfire occurs, the fire could be managed and the fuels and vegetation kept at sustainable levels. In that way uncontrollable wildfires would be avoided.

For example, in landscapes with chaparral and brush, we prescribe a hotter burn, one that will consume the brush, since this is the type of fire needed to meet this goal. In forested landscapes where we have pine needles and bear clover fuels on the ground, a slower, cooler burn is needed. We tend to start these types of burns from the top of the slope and let it back down slowly. That way we get the benefit of the fuels on the ground being consumed and less chance of the larger fuels—the trees and most of the brush—being damaged. If we were to light the fire from the bottom of the slope we would not get the same results. There would be too great a potential to lose some of the vegetation we want to keep. We want that slow fire to help us keep the flame lengths low so the fire can consume the fuels and increase the reduction of the fuels.

In prescribed burning of the oily, flammable groundcover like bear clover, also known as mountain misery, higher intensity fires are not uncommon. *Courtesy of Burt Stalter.*

"Sometimes it can be difficult to use prescribed fire though," said Burt. "Many people only see the pictures of devastating wildfires where there is nothing left on the landscape and believe that is what happens every time there is a fire. Many folks are not familiar with what it means to be in a fire adapted landscape and that not all fires are bad."

*Let's say we're in what in the past was a ponderosa pine stand, but over many years of having no fire in it, the stand is now mixed with cedar and fir. We want to try and reduce the amount of cedar and fir to allow the ponderosa pine to grow better and be healthier. But this "forest" of ponderosa pine with a thick growth of cedar and fir is what the public has become used to seeing over many generations and believes is what a forest is supposed to look like. They don't want the trees burned.*

*But with wildfire in the summertime, all those small cedars and firs are problems for us. Those trees don't belong there. They were never there to begin with. They burn really hot. People don't see the issues those small dense stands of trees are creating, firewise. They compete heavily with each other for resources such as water and nutrients and crowd out the few sugar and ponderosa pine that do become established. These dense tree stands generally contribute to unhealthy forest conditions.*

*I feel one way of helping the public to understand the value of a forest is through education.*

"We try to educate the public all the time, at campground programs, nature walks, all of those things," said Richard Bagley, manager of forestry operations at Shaver Lake for Southern California Edison.

*Up at Shaver, we find that people like the Central Park look, you know, grassy and nice trees, everything clean and groomed. But that's not what a squirrel likes. That is not what a coyote likes or a mountain lion. The animals' view of a good place to live is different from the human idea. So by teaching the people to look at it from a different point of view, I mean, what is it that a bear likes? Besides your garbage, that is.*

To get to that natural, healthy forest or to manage by fire for any purpose involves long-range planning. And that might be when the work becomes nerve-wracking, as Biswell described it. Many things can prevent a planned burn from taking place. A change in weather, a sudden downgrade in air quality, a lack of resources because of wildfire demands—any number of things could dash the plans of a fire manager.

"Even knowing things might change, we plan prescribed burns way in advance," explained Richard.

*We turn in our plans to Cal Fire and the air district ahead of time, lay out contingencies, look at the public health issues, things like that. We target our prescriptions for when temperatures are lower and the fuels are moister. It's more controllable during those times, and a lot less smoke is produced than would be experienced during a wildfire. The federal agencies have even more opportunity to adapt. Let's say a lightning fire starts in that prescribed spot during the hot summer months. You adapt. It's not an automatic let-burn. They are able to go in and manage it. You try to allow the fire to act in a natural way. Let it burn to the river or to the rocks or to a hiking trail it won't jump.*

*Managing fire could be as simple as just watching. It's adaptive. If the fire is surrounded by rock, they may just fly over and make sure there's no place for it to go. They wouldn't even put anybody on it, if that were the*

Managed wildfire, lightning origin. It is allowed to burn under watchful eyes instead of suppressing, which is good for wilderness areas where the terrain keeps the fire contained. *Courtesy of Mike Esposito, fire ecologist, Southern California Edison Forestry.*

*case. But if the fire starts burning too hot, or the winds come up unexpectedly and the fire begins behaving erratically, they have to adapt again, change directions. In that case, they would send in the crews to build line. And, of course, if it turns into a wildfire, it's an all-out effort to stop the fire as quickly as possible, and that acreage will get a different plan.*

For Carolyn Ballard, fire management officer at the High Sierra Ranger District of the Sierra National Forest, taking on an adaptive management approach, one that involves observing, learning by doing and incorporating new information into the management plan, has allowed her to ask questions and get her answers out in the field.

*One of the questions we asked was related to the Pacific fisher. They didn't want us to do any prescribed burning anywhere within fisher denning activities. But with the research we've had going on, we've learned from the sensors about their denning activities, their natal den versus the birthing den and when they begin to move. That tells us at what periods we can begin to do prescribed burns that won't put those natal dens at risk. Before, they told us no burning until after the end of July. But now that we know more, we're able to adapt our burn windows to mid-May when those fishers are starting*

Prescribed burning is planned for the cooler months when the chance for the fire breaking out and the problems with air pollution are reduced. *Courtesy of Burt Stalter.*

> *to move their young around. That gives us two extra months of ideal cooler weather to plan our prescribed burns. We've also put temperature sensors outside and inside their dens once they've vacated them and charted the temperatures and carbon monoxide levels during a prescribed burn to know if we're imposing a risk on the fisher. So we're learning quite a bit.*

Carolyn admits that sometimes adaptive management means setting priorities for competing needs.

> *We make these decisions as an interdisciplinary team: a silviculturist, the one who looks at how well trees grow and how they survive; a wildlife biologist, because we have so much habitat and so much variety—fishers, owls, bats. We have a hydrologist that looks at potential for soil movement or soil compaction; an archaeologist, because we have cultural resources that need to be protected. We have a botanist, generally, because we have to consider endangered plant species, and we have an aquatics person who works with us on the frogs and toads and things like that. So that's how we determine what goes on in the landscape, what we can burn here, what*

*needs to be protected, what's more in danger if we don't do anything and then we lose all the habitat.*

*Like through our fisher research, we've learned that the fisher likes to use snags and likes to feel protected within a grouping of trees. They like it dense. They like ladder fuels. They like to have ground cover so they're not preyed upon. Plus, it keeps things cooler. But those are also the kinds of forest structure that might more readily burn. So we have to find a balance of all these different niches for wildlife as opposed to what may or may not survive in a wildfire, or what would be ideal for growing big trees. We still like to grow big trees because there are a lot fewer on the landscape than there used to be.*

*This fire season, 2014, is going to be a real challenge. Here it is January already and we haven't had a drop of rain. This is the driest year on record in 120 years. So we're almost charting new territory. Here in the Service, we're already talking about the fire potential when fire season comes. What it will be like. We would have to have rain and snow every four days all the way through April and into May to get to a normal seasonal snowpack, to get enough moisture in our 10,000-hour fuels, our big downed logs, which are bone dry right now. If they don't get moisture they'll burn rapidly.*

*Fuels gain moisture, more rapidly or slowly depending on their size. So a one-hour piece of wood can change or fluctuate in its moisture uptake or in how much it loses in an hour. That's pine needles and grass and things like that that. Air humidity can make them moist or they can lose it in an hour if you get a warm, sunny day. Ten-hour fuels, those are a quarter of an inch in size, the small sticks and branches. Then there are the hundred-hour fuels, which are the one-inch ones, and the thousand-hour fuels, which are three inches. Those are the surface fuels, a small tree that's three inches, or a branch, for example. Those take a thousand hours of moisture to become fully moist. Where fire won't burn it. But it also takes a thousand hours of dryness for it to become fully dry. When it can be fully consumed by fire. A thousand hours is forty-two days, a little over a month. We've had no rain now for almost three months. Can you imagine how dry those fuels are even now?*

*Setting priorities for prescribed burning in a serious drought year presents many problems. Way up at eight thousand feet, up in a redwood forest, you might only get fire every thirty years, maybe every hundred years. It's generally too moist by the time the snow melts. It might be snow-free only between August and September and then get snowed in again. It's too wet to burn most of the time. That's why the fire interval is so long. In years like this, however, who knows? That plays into how fire-adapted those different ecosystems are.*

*We know in our low elevations the ponderosa pine is fire adapted, and it's a species that needs fire to regenerate and stay healthy. If I have projects at both the high and low elevations, where am I going to go to put fire? I can only put fire into one spot, which one will I do? We tend to go to the lower elevation first because historically that had the most frequent fire, so it has missed the fire in there that keeps forest open and the nitrogen cycling and the plants re-sprouting. More often, fire has been missed, so it'll need more fire, so if we are going to use our manpower and all, we'll choose that lower elevation project.*

"For the fire manager, fire is only one element contained in a prescriptive plan. Forest restoration involves managing fuels with every tool we have," Burt said.

*We have fire. But we also have other things we can do to help restore the forest. Mastication, for example. Mastication has been around for a long time. It was done during the '70s and became commonplace in the early 2000s. During that time, the National Fire Plan reallocated funding towards fuels management and reduction on federal lands.*

*Mastication was seen as a treatment solution to fuels management. Mastication treated areas have been a double-edged sword, however, when wildfires have burned into the treated areas. On one side, the fire behavior was changed from one of very high crown fire type flames down to lower surface fire type flame lengths, but with very high concentrated heat outputs. This has made initial attack in these areas more difficult because more water has to be used to cool down the fire before the crews can construct fire lines.*

*The Carstens wildfire near Mariposa Pines in June of 2013 was one of the first wildfires on the Sierra National Forest that burned through large old brush fields that had been masticated over the previous ten years. Most of the treated areas burned with high severity fire effects. However, the overall consensus was that the mastication treatments worked as had been planned because they contributed to the ability of suppression resources to contain the fire quickly and protect nearby private property.*

*What we learned from this fire was that mastication can be a very useful tool if approached in a multiple-stage prescription. After mastication, we need to reduce the fuel bed with prescribed fire to increase the effectiveness of the initial treatment.*

*The timing of the prescribed burning is the key to each approach. In the more traditional method, you go in, masticate down the brush and fuel ladders and put fire on the ground to clean up the masticated slash. Another approach is to burn the surface fuels first, masticate the dead tree and brush skeletons down and then come in with a final entry of prescribed burning of the masticated slash. This last method saves having to do three or more entries with fire since the mechanical treatment does what the second fire entry is usually planned for. The end result is, I don't put up as much smoke in burn entries, which is important for smoke management within the San Joaquin Valley airshed.*

Richard describes mastication as one of the surrogates for fire that accomplishes only a part of what fire does.

*Mastication doesn't replace fire because it can't do the ecological work fire does. It does reduce the fuel loading and helps shift the forest plant mix to more heterogeneity, but it doesn't promote those plants that need fire to reproduce.*

*There are a bunch of things our non-fire approaches don't achieve, like scarifying seed—opening dormant seed in preparation for germination—enriching soils through rapid nutrient recycling, enhancing fire dependent plant species and the wildlife that depend on those plants, all those parts of the forest system that are nurtured by fire. We would do a lot more prescriptive burning if we could.*

Logging is perhaps one of the more controversial activities in present-day forest adaptive management. People have been sensitized to the past practices of clear-cutting and road-building to accommodate logging that took place in the national forests until the mid-'90s. And there are those who simply do not want any trees removed from the forest, the environmentalists and preservationists who began challenging the Forest Service logging practices in the wake of environmental laws such as the National Environmental Policy Act (1969), the National Forest Management Act (1976) and the Endangered Species Act (1973). Protection of American Forests became a national movement and ultimately influenced changes in management's focus toward forest preservation and protection. The struggle to eliminate tree removal continues, whether for salvage after a wildfire, fuel reduction or timber sale.

"Here at Edison," explained Richard, "when our people go out to mark the trees for removal, they look down the road to envision what they want to achieve with their treatment."

*They're like wood carvers. If you give a wood carver a block of wood, they have in mind what they want it to be when they are done. They might even tell you that the piece of wood tells you what it was meant to be.*

*When you look at the forest, you don't ignore what's existing or what's going on there. For instance, there is a bird's nest in that tree. It might be the cedar, and you would prefer to take it out. But if there's a bird's nest in that tree, you don't take it out. Or if it might be a fisher den, you don't take the tree out.*

*There are hundreds of different things the forest is telling you about what needs to be done. Not for just right after a timber sale, but for a hundred years from now. That's what you're picturing. So just like a wood carver, we don't go and in one treatment take everything out. We take our time, carving in one area and going back and carving in another, and keep observing what's going on. We're going to do it every ten years or so, and that allows some recovery time for the plants and the animals. And then we treat it again, only taking part of it.*

*The idea is that maybe eight, ten treatments out, we're going to be pretty close to where the forest needs to be, as opposed to saying that in this acre there should be thirty pine trees, ten fir, twenty cedars, sixteen oaks and go in at one time and make it that way. That would be very disruptive to wildlife and habitats and very hard on the ecosystem to try to do that.*

Following are descriptions of three ongoing projects being conducted on the Sierra National Forest that are using prescribed fire. The first is a prescription to encourage tree growth in the Nelder Grove of giant sequoias. The second is a habitat improvement project on Kinsman Flat. The third is a forest restoration project at Huntington Lake.

Nelder Grove as described by project director Mark Smith, battalion chief in the Bass Lake Ranger District.

*This area has been closed for any type of management except for recreation. Only lately, because of the heavy concentration of fuel on the ground, which needs to be cleared if only to protect the standing trees there, we're trying to reintroduce fire to get some new trees to grow and to increase the overall health of the stand.*

*There's a lot of work being done to figure out how to bring fire to redwood groves to bring back redwood health and to encourage some new tree growth. We have a tree up there that's twenty-two feet across. That's something you're going to want to protect. In those areas that are congressionally protected, not just forest protected, you're very limited on the type of work you can do in there, so fire has always been excluded, but not only that, you can't log in there either.*

*So there really hasn't been any type of disturbance in those forests, anything that has cleaned them out. So what has happened is that all the fir trees that have grown up under there have died and fallen. Plus, two years before the drought we had some heavy winters, so there's been a lot of snow kill, heavy winds, blowdown. You have nature putting a lot of fuel on the ground and that's the current situation.*

*Because of the historic story from the logging of the giant sequoias, there are a lot of stumps in there also. So if a fire started, it would burn all that up, but it would also burn up all the stumps. Naturally that wouldn't be a problem because they're already dead trees, but the issue is, if we're preserving them for a historic value, then now we have to go in there and do some type of management.*

*What we're trying to do is hand pile the heavier fuels and burn it. But also, knowing that the sequoias need fire for production to open up the seed cone, the thing is to let the fire move around in a very controlled environment and try to encourage some new growth in that area. We went and looked around and saw that we did have a small success.*

*We let a little piece burn up at the Nelder Grove, probably fifty by fifty feet. We found two new trees growing. Two is not much, but it shows that the technology of introducing fire is working.*

*It took me eleven years to get permission to start this project. This is 2014, so I'm about a year and a half into it. It took that long because of the environmental protection laws. It's also a denning site for fishers, a spotted owl habitat and a goshawk habitat, and there's a big area of archaeological resources. They were afraid we'd burn up the archaeological resources. What they're seeing now is that if we don't clean it up and get all that debris out of there, they're going to lose the archaeological resources anyway.*

## Kinsman Flat as described by project leader Burt Stalter:

*The Source/Kinsman is a multi-objective project, with wildlife habitat improvement and cultural resource improvement being the main emphasis. The wildlife we are focused on are deer and wild turkey. This area is key winter deer range for the San Joaquin River deer herd, who use the high concentration of chaparral and acorn-producing black oak trees for food and cover.*

*This resource rich area is also home to the local Mono tribe who, to this very day, gather foodstuffs and resources for baskets and ceremonial purposes on this land.*

*Current research has found that chaparral fuel beds burn on a longer fire rotation than was once thought. It doesn't burn at the same fire return interval as the Mixed Conifer or Ponderosa Pine forests. We treat it like an old growth lodgepole forest. It's a long-term rotation and when it burns, it burns completely down, real clean and then it takes time to regenerate itself. The recent Aspen (2013) and French (2014) fires on the Sierra National Forest burned through many of these old shrub fields along the San Joaquin River Canyon, and this was actually a good fire for these types of fuels and for the wildlife that will have a lot of young-aged browse for food as these plants recover. The type of vegetation in the Source/Kinsman area I'm dealing with are the eighty- to one-hundred-year-old chaparral fields. This age of brush is difficult for the deer to walk through and not nearly as appealing for them to eat. That is why I'm trying to get this brush burned down to let the deer go in there and make it through the brush, plus it will give them new brush shoots for them to feed on.*

Burning manzanita for deer habitat. *Courtesy of Burt Stalter.*

*High intensity fire is needed to do that, and I use different methods to safely produce that high intensity fire. On steeper, inaccessible terrain I use aerial ignition devices such as a helitorch. The helitorch is like an aerial driptorch. It has a barrel suspended below the helicopter, with a fuel mixture of an additive called Flash 21 and gasoline. This additive gelatizes the gasoline to a pudding-like consistency and drips out of the barrel. There's an igniter that sets this dripping fuel on fire and the lighted gel drips down onto the brush. That's all controlled from inside the helicopter. The pilot uses what is called a firing pattern that is planned before the helicopter leaves the ground so he doesn't light in an undesirable way that would create too much fire, causing undesirable resource effects. The fire is started in a dot-like pattern across the slope, so it can make short hot runs up the slope to create the fire needed to burn holes into the brush fields. Large old brush fields usually are lacking in ground fuels so slope and wind are needed to get fire to burn these types of standing fuels. This type of ignition is cheaper and safer for the crews because they do not have to try to fight their way through these brush fields with handheld drip torches.*

Helitorch with flames preparing to drop fire into stand of knobcone pine. *Courtesy of Burt Stalter.*

Helitorch dropping flames into stand of knobcone pine. *Courtesy of Burt Stalter.*

Flames on the ground among the knobcone pine. *Courtesy of Burt Statler.*

*On the flatter terrain, we're starting to go in with bulldozers that drag a large ship-anchor chain to crush the brush and cut pathways through the chaparral. Then we can go in during the wintertime and burn the slash with hand-held drip torches.*

*The reason for using crushing, from a prescribed fire standpoint, is it provides greater flexibility in utilizing the very limited burn windows and air pollution control district approvals we get to accomplish our burns. To burn standing manzanita and standing chaparral, you need some fairly high wind conditions, and it has to be dry. Moderate environmental conditions usually are not enough to get it to burn. So we can't do it during our normal prescribed burn windows, which is when fuels are not as dry and wind speeds are not as high. This area of the San Joaquin River canyon and Lions Point produces very complex wind patterns that create conditions that are more difficult to safely burn under. The past fire managers who worked out here beginning in the mid-1950s used this crushing technique for the same reasons. I figured they also felt the risk was too high for a burn to escape without crushing the brush first.*

*The crushing and the way we're manipulating the brush for deer habitat has been going on since 1955 in this area of Kinsman and Lions Point. In the early '80s, they repeated the crushing. I'm doing it for the third time. So every thirty years or so is the return interval where we go in and reset the brush to a younger aged class using this crush and burn technique. The original research project was very large back then. They covered approximately eight thousand acres of brush crushing and burning. I'm not dealing with anything that large with the Source/Kinsman project. Back then, using the crushing and then burning it, worked. I'm not going to think I'm smart enough to improve on that. They had it right, so why change a good thing?*

Huntington Lake, as described by project leader Carolyn Ballard:

*The Huntington Lake project is a forest restoration project, not just a prescribed burn. We will be using all of the tools in our toolbox. In fact prescribed fire is very difficult in this basin with all the structures—so we have to limit where, how much and when we burn and at what intensity. The fire would have to be low intensity. This is crucial so as to protect the structures.*

*We're pretty worried about the situation up there. Our red fir forests are starting to die. We think part of it is ozone. Part is climate change. We used to have fire season end in September. We used to get a lot of rain and snow in September. Now we're seeing October and November. So we're seeing issues with tree mortality at the red fir levels, which is 7,200 feet elevation. A lot of those trees are starting to fall.*

*There are a lot of cabins around there, structures that are owned by individuals, but who have special use arrangements with the Forest Service. The trees are putting the cabins at risk. Also, when a red fir comes down it adds a lot of fuel loading on the ground, a lot of slash. The potential for wildfire, particularly within that urban intermix around the cabin tracts, is very high in some places. So we'd like to be able to go in and thin out some of the trees.*

*That was an area that would have burned every fifteen to thirty years, at least every hundred years. But we've completely fought and suppressed fires in that area, so none of the ground fuels have been getting cleaned up. We have a lot of white fir coming in underneath and a lot of lodgepole pines, and some of the stands are getting very, very dense between the cabin tracts. So we'd like to go clean those out, get some of the slash out of there, clear out the duff, thin the stands a little so they're a little more fire resilient. We don't want their cabins to burn, and we don't want trees to fall on their cabins.*

*But some of them won't agree to let us do it. People say, my grandpa planted this tree. You can't take it down. Or, don't touch this area, this is wilderness. There are four hundred fifty structures in there. That's not wilderness. Basically, some don't want us to do anything. And then there are those who tell us, you need to get in here and do something now. But we can't go in there mid-summer to do the work. It's a major recreation area with a lot of summer usage.*

*We had a lot of resistance and a lot of people complained, that is, until the Aspen fire started* [in the summer of 2013] *and now it's like, it's coming over the mountain and when are you going to do something? How fast can you get in here? When you do, we don't want all of our trees taken. So we're beginning to see a bigger picture way of thinking, more long-term.*

*We do education. Some people want to learn and others don't. We have public meetings, and some people turn out and others don't. The Huntington homeowners have very active association, and we have a good working relationship with them. It's like working with a collaborative, trying to find a balance, coming to some consensus about what is the right thing to do and at what level.*

*We try to come to it from a scientific point of view. As land managers we want to protect the basin. It needs a lot of work because we're losing a lot*

Prescribed burning directly under a tree: if you want the tree to survive, back the fire away at low intensity. *Courtesy of Southern California Edison Forestry.*

*of the big trees that are being outcompeted for moisture by the smaller ones coming in. We've been talking about doing something at Huntington for seven or eight years, but it's going to be a slow process because we can't get in there until after Labor Day. So we're moving through that environmental process and we're trying to balance what we're doing in the Huntington with everything else we have to do. But, really, even if we just chip away at it little by little, at least we'll be doing something.*

*Part III*

# LIVING WITH FIRE

Experts tell us that wildfires are becoming more frequent and more intense. They say it's not a matter of if but when the fire will come. So in April each year, we begin to anticipate the arrival of the fire season, almost expecting disastrous wildfires to happen. In early summer, as the air heats up—particularly in a year when the snowpack is down 50 percent and it hasn't rained in three months—we begin watching for the telltale sky, when the sun will turn orange behind a shroud of brown haze. We wake up in the middle of the night and look out our windows, sniff the air. Has a fire started somewhere as we slept? Fire talk has entered our conversations. In coffee shops, at the supermarket, we chat facilely about fuel loading and suppression, backfiring and cutting line. And for the next three months, we will hear about fires happening somewhere nearby, sometimes in our own neighborhoods.

# 6

# MON DIEU!

## THE FRENCH FIRE

The Sierra Vista Scenic Byway is a ninety-mile road that circles through the Sierra National Forest. On a trip along the byway, visitors travel through abundant forests. They have the chance to stop at historic sites, climb domes, hike, swim in creeks and enjoy vistas of the stunning Sierra Nevada Mountain Range. A popular side trip from the byway to Mile-High Vista offers views of distant peaks, including the Minarets, Mount Ritter at thirteen thousand feet and the famous ski resort Mammoth Mountain. Two thousand feet below the vista, the Mammoth Pool Reservoir and the San Joaquin River canyon lay surrounded by the towering granite domes of the Sierra National Forest.

Since July 2014, a tour of the byway includes a fourteen-mile swath along Road 81 of forestlands devastated by a wildfire that began at an abandoned campfire on a bluff overlooking the San Joaquin River. By 5:45 a.m. on July 28, the fire from the campsite had worked its way through the pine needles and down into the San Joaquin River drainage at Rock Creek. There, winds drove it north through the river canyon toward Mammoth Pool Reservoir. The view from the Mile-High Vista, itself ravaged by the fire, now contains stark evidence of the fire's movement. "All the trees on Mile-High are dead," said Burt Stalter. "It's hardly recognizable."

If things had gone along normally during the first days, the fire would probably not have climbed up over the Mile-High Vista nor crossed Road 81 and traveled another two miles before firefighters could stop it. It would have continued its push north, where the natural winds in the area of the

Mammoth Pools Reservoir seen from Mile-High Lookout, which was ravaged during the French fire. *Collection of the author.*

reservoir would have shoved it into the Chiquito Creek drainage. From there, the fire would have traveled to the rocky edge of Yosemite National Park, at which point the firefighters could have driven it into the rocks and checked its advance.

But things did not go normally. On the second day of the fire, a cold front passed through, setting off a wind event that turned the fire south and sent it up out of the Rock Creek drainage.

"We could see the fire down below at the bottom of Rock Creek, and it was coming up towards us," said Burt.

> *We had pushed lines out on the ridge to the east of us. We had a dozer pushing out that way. We had a crew that had started backfiring that ridge. Our plan was to bring the fire up towards Forest Road 81 and start backfiring there, to punch the steam out of it and keep it down in the hole.*
>
> *And it was working fine. We had retardant down there. We figured we could hang it up at the top of this ridge and keep it forced into the*

*river canyon and not let it come up Rock Creek. That's what we were hoping. But when that wind shift happened, it took all that fire and turned it and ran it up this way. We weren't anticipating that big of a wind shift that early. It happened at six or seven in the evening. The wind shift normally happens around nine or ten at night, when it's cool.*

*So, as soon as the crew was starting their backfire, the fire came up out of there, just one big push. It came up faster than we could get fire on top of it. We had to pull everyone out of there. We were able to get the dozer out of there just when the fire started crowning.*

*It was just a big wall of flame coming up at us, and once it came up out of that drainage, it just kept going, and the crews got into defensive mode. There was the Hogue Ranch and the Ross Cabin to consider. They wrapped the Ross Cabin and planned to defend the ranch with anything they had. Luckily the fire never reached there.*

In post-fire scouting, Burt was able to intuit what had happened that day when the wind shifted and what happened after that day.

*The fire had started in the bottom at Rock Creek and it was coming up the drainage. Our first line of attack was at the southern end of the fire down in the canyon. We had put in a contingency line the year before for the Aspen fire, which happened directly across the San Joaquin. We had built the line along the ridge and brought it all the way down to the river—about a mile of line. So when this French fire broke out, that line was there. The crews didn't have to spend any time on it. It was really clean because we hadn't rehabbed it the year before, just left it as a fire break. And I'm glad we did because it turned out to be a really important line. It saved days and days of work. They just went in and started backfiring off of it. They had the southern end of the fire stopped within two days.*

*But then that weird event happened. The wind wrapped around the canyon and took the fire with it and kind of forced it up the slope. You can see that pattern distinctly, all the burnt trees where it really roared out of there. You can see the high intensity runs as it climbed up over there and it just kept running to the south. That's the first run it did that night.*

*By day three the fire crested. It came around and threw spots over the top and started burning downward. At the slope transition, it was a backing fire going down. It burned up over that slope with a real hot*

Severe erosion after unnatural wildfire and winter precipitation. Stream and river habitat downstream was destroyed. *Courtesy of Southern California Edison Forestry.*

One year after unnatural wildfire, the area is devoid of vegetation, except for a small amount of brush. *Courtesy of Southern California Edison Forestry.*

*head fire and that's where it came up over the Mile High Vista. It was all bear clover and heavy dense trees down below, so it got a big head of steam and headed up.*

*When the last big run happened, we were under an inversion from the smoke that helped check the fire and kept it from making huge runs every single day. It was stuck in the inversion for a couple of days. It would kind of chunk its way through, making little runs here and there. Then the smoke cleared out of the basin and it got really open, that's when it came to life and made big, big runs. It was like the lid was taken off the top. That allowed it to grow vertically. It ran about four thousand acres before the weather moderated. There was some cloud cover, and it moistened up a bit. Then we were able to cut a line along Shuteye Ridge.*

"Building fire lines is the key to firefighting, and ridges are the first choice of where to build them," Burt explained. "That's where the wind peaks out, and it makes a strong spot for us to do backfiring and letting the wind push the fire back on itself. Anywhere mid-slope we couldn't hold it. And you don't want to do it on the bottom of the ridge and backfire up, not if you can help it. You want to backfire on the top of the ridge and let it back down. It's a slower fire."

*The tools for building line are pretty basic, and they're not likely to change very much: hand tools, chain saws, bulldozers. You can have all kinds of air attack with retardant and water drops. It's all important. But it has to be backed up by the crews on the ground.*

*There are two types of crews: type one are the hotshots and type two are the on-call firefighters. The hotshots have special training. They can break up into small groups, do burnouts, a lot of things. They can multitask, act as dozer bosses, even cut dozer line if needed. They have that ability to make decisions. And that's a really valuable asset during fires because, as an incident commander, if I have a fire, especially during an initial attack, which is the most critical time, if I bring in a hotshot crew and I have a dozer available but need someone to go scout it out, I can use the hotshot crew to do that. They are self-sufficient. They can pretty much take care of their own needs for forty-eight hours—foodwise, waterwise, everything. They can camp out at the fire. In fact, they would rather do that because that's when they are most productive.*

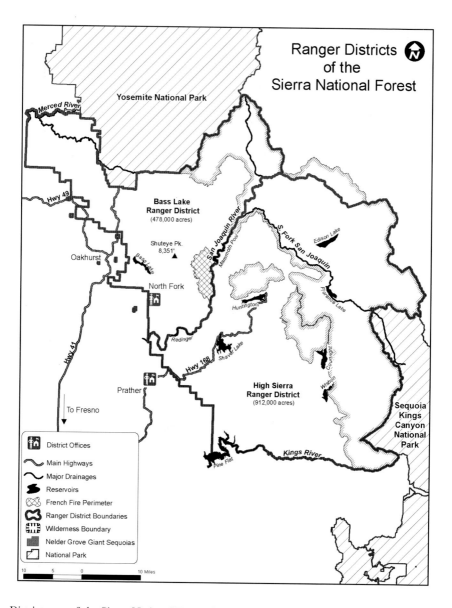

District map of the Sierra National Forest. *Courtesy of Heather Taylor, former firefighter, now a Type 1 team as map person.*

So, then the question arises, what do you do to manage fourteen thousand acres of dead trees and brush and hundreds of acres of scorched land? Burt, who, as fuels manager for the Bass Lake District, will oversee the restoration of the forest, described the project as daunting.

# French Fire

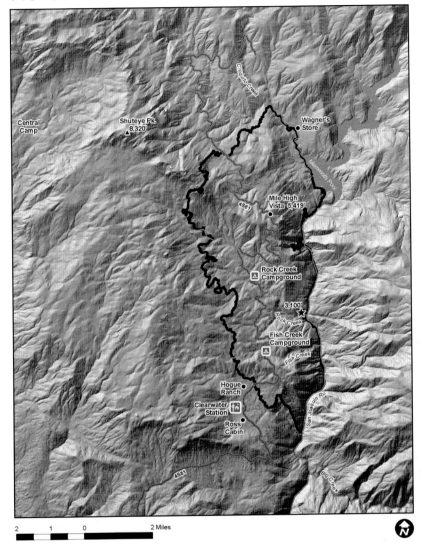

Map of the French fire. *Courtesy of Heather Taylor, former firefighter, now a Type 1 team as map person.*

*I've been here for twenty-five years and nothing like this has ever happened in this area. To drive through the woods and see so much of the forest burned is, like, wow. Am I going to have to drive through a burned up forest for the rest of my career?*

The fire line cut along the top of a ridge. *Courtesy of Burt Stalter.*

*It's kind of humbling, you know. You have certain trees that you've become familiar with over time, ones that you recognize. This one has a little conk; this one has an odd tilt—little things like that. You get used to it over the years. And then, when fires come like this and you drive through, it kind of resets your whole view. It's a whole new forest.*

*And it's a whole new forest for other reasons, too. Because of this one hundred year history of no fires, and top it off with the drought, we're seeing things we've never seen before. The biggest trees, like the ten thousand–hour fuels, are generally the indicators of the severity of the drought condition. Normally, you don't see those things burn to white ash. They hold their moisture for longer periods of time. Now we're seeing them burn to white ash.*

*Also, we're coming across trees that burned deeper and further than anyone had ever experienced. Twenty feet away, you're falling into a stump hole because the fire burned that far into the root structure because it was so dry from the drought. Take some of the large green pines. Normally, those green trees are not damaged very much by fire. But now we're having some of those fall over because the fire was so hot it burned down into the green roots and burned them out. No roots, and the trees fall over.*

*The first consideration for restoration is getting rid of the hazards. Lots of trees are in danger of falling on the road or being blown down by the Mono winds in this area. This part of the district is so wind-prone that if we don't do some sort of hazard tree reduction, our road will be clogged up with trees and debris, just from the wind knocking down the vulnerable trees.*

*Then there's replanting. We'll do it, but with ecologic considerations. It's different now than it was back then. When I started with the Forest Service, it was basically, "Put a tree in every open space we can find." That's because we were basically planting timber, growing wood fiber. Now it has transitioned into a different paradigm. When we plan the reforestation, we do it as a team. We consider wildlife habitat, stream protection, aquatic species, cultural issues. We're trying to look at it from the standpoint of putting a forest in that will resist fire. It'll grow up, and we can keep it relatively safe from wildfire by going in every twenty years or so and clear it out with prescribed fire. After all, this French fire has reset nature, and we should take advantage of this opportunity to start over. All I want to do is leave a green, healthy forest for the public to enjoy.*

# ON ANOTHER NOTE

## Air Patrol: Then and Now

*I don't fly the plane. I just sit and look out the window. It's the greatest job in the world.*
—Patrick Smiley Tierney

From the very beginning, spotting was considered an important first step in the management of wildfire. In the Sierra, Wofford's order to "get 'em early when they're small" prevailed. Charles Shinn romanticized the lookout life. Rangers clambered to high places in search of smokes. They built roads and crude lookouts and patrolled the hinterlands on horseback. Finally, in 1915, reports Gene Rose, "a weary firefighter, S.L. Cotta, a ranger on the Eldorado National Forest, looked to the heavens and then asked if the service had ever investigated the use of airplanes or dirigibles to spot forest fires." It was an idea that attracted Coert DuBois at the time, but World War I intervened.

In 1919, after the armistice, with the urging of Coert DuBois—having returned from the army as a major general—and in collaboration with the

Army Air Service, the Forest Service introduced aerial spotting in California as a potential tactic for improving response time to fires. Although the project was abandoned after a few years because of budgetary, personnel, logistic and equipment issues, the short-lived experiment was to have a far-reaching influence on the development of aerial support in fighting fires.

In his memoir, Roy Boothe writes about the brief era in California when airplanes were given a trial for fighting fire. He tells of his personal impressions as an aerial liaison officer.

> *In most instances, the pilots were equipped with "Jennys", (the old training ships used by the Army Air Service), which proved to be satisfactory so long as the altitude of the area to be patrolled was not too high. The ceiling of this type* [of] *plane would not permit it to be flown with safety over rugged terrain of high altitudes, however, and they had a very short gliding angle so that if anything happened to cause the motor to cut out the pilot had to select a landing place within a comparatively small radius.* [The Jennys were replaced after several months by the DeHavilland DH4.]
>
> *During this trial flight work, a couple of officers and ships were assigned to a small base near Fresno,* [Chandler Field] *and Supervisor* [Maurice] *Benedict and Ranger Mal McLeod were* [flown] *over the Sierra Forest by the pilots to get an idea of the fine view to be obtained from the forest from an airplane. I happened to go out to this landing field early in September, 1919, and the pilot in charge noticed my Forest Service uniform clothes, and asked me where I was stationed. On being informed that I was District Ranger in charge of the Kings River District on the Sierra National Forest, he asked me if I had been up in a plane, and on being informed that I had not, he invited me to take a trip over my district...and so without much thought and rather with surprise, I found myself in a flying suit, taking off on my first airplane trip. It proved to be an exhilarating experience, and I was very much surprised and pleased with the wonderful view that one could get of the country from an airplane.*

A year later, a training school for pilots, observers, radiomen and Forest Service liaison personnel was established at March Field in Southern California. Boothe was invited as one of ten rangers from three northern regions to attend the training.

> *I was assigned as Liaison officer and acted as observation officer in making flights over the North and South patrols, acquainting each of the pilots with*

*as many physical features and names as I could, as we flew the Southerly course over the Sierras to Bakersfield, and the Northerly course as far as Sacramento. I was stationed at Fresno for about six weeks.*

Boothe found the airplanes to have some advantage over the lookouts in that they had a much better view of the country to be patrolled, particularly at times when heavy haze or smoke drifted in to restrict the vision of the lookout men.

*Haze or smoke builds up more or less in layers over the country and usually the canyons are filled with a fairly thick layer through which the look-out men cannot see to good advantage. The observer from the plane, however, flying over the top of this haze, has the advantage of looking down through a comparatively thin layer, and therefore he has little difficulty in locating any smoke or other object that he might be in search of.*

*The disadvantage of the airplane, of course, was the fact that after the patrol was made a fire could start within a few minutes and no opportunity would be given for the air patrolman to pick it up until his next flight, which might be several hours or even a whole day later, whereas the look-out man being on duty continually had a much great[er] chance of picking up a smoke soon after its inception than would the airplane patrol.*

*Radio instruments at that time had not been developed to a satisfactory degree of performance, and considerable difficulty was also encountered in either sending or receiving reports by radio from the plane. I carried on some experiments in dropping messages with a fair degree of success, and I believe that others acting as observers also resorted to dropping messages when they wanted to report fires.*

*After these two seasons of experimental flying, it was finally decided that the only practical value of air patrols that would justify the heavy expense of operating them was Reconnaissance flights over going fires for the purpose of determining the actual conditions of the fire and for a quick examination of a going fire by some officer directly responsible for its control, in order to give him a bird's-eye view of all sides of the fire so that plans for its attack and control could be made more intelligently.*

What Boothe did not, and probably could not imagine, was aerial firefighting with water and retardant and reconnaissance that coordinated on-the-ground and aerial attacks. It's what Smiley Tierney does. It's called Air Tactical Group (AirTac). As Smiley explained:

*If you're working AirTac, say at Fresno, we'd be the first ones at the fire. We come to the fire and our job is to fly over and evaluate it. We call back to the dispatch office. We tell them where it is, how big it is. We direct the crews from the plane. We can see them down there. A lot of them carry mirrors* [reminiscent of the heliographs described by Charles Shinn in 1907]. *And you'd be amazed how that can stand out in the middle of a forest. You can see them from 40 miles away. That is one of the most effective tools they've got.*

*We'll call in the air tankers that drop the retardant and the helicopters that'll drop the water. The helicopters work at 500 feet. We stay 500 to 1,000 feet. If we call air tankers, the air tankers can go down to 150 feet to do their drop. So part of my job is to coordinate all of these aerial resources and keep in contact with the people on the ground, because if something is going wrong, we'll be the ones who can see it. When you're down there and it's thick smoke, you have no idea what's going on. Your whole world cuts down to 40 or 50 feet. So the AirTac's job is to provide the eyes in the sky. We're kind of their safety net.*

*It's hard to describe how chaotic it can get in the cockpit because you'll have people in the air tankers wanting to talk with you, the helicopters want to talk to you, the guys on the ground want to talk to you, the dispatch office wants to talk to you. And then, there might be someone who wants to pass through the airspace. At that time, we own the airspace, so they have to call us. Technically, it's a Temporary Flight Restriction (TFR). It's like when the president of the United States is flying, you're not supposed to fly in that space. So for example, if during a fire, you're flying your plane and you want to land at the Mariposa airport, AirTac has to make arrangements for you to get through. We have to coordinate all that so you don't get run over by a DC10 that was converted from a passenger plane to a retardant-dropping air tanker.*

*I flew on the Rim Fire in Yosemite in 2013. That was chaotic, sitting there trying to decide who's the most important person to respond to. I tell the others to stand by. "Stand by dispatch, stand by helicopters, stand by air attack. Tanker one zero you're cleared in. Division alpha, tanker one zero's coming in. Dispatch a load and return for tanker one zero." If you sit there and start having conversations, you're lost.*

*I didn't start out training for this job. It kind of came my way. In 1987, I was falling a tree down at Los Padres National Forest on a fire, and one of the branches came down and kind of center punched me into the ground. I was hit in the back and was laid up for a while. Then a crusty fire manager looked at me and said, "Don't think you're going to stay home. Why don't you go down to the dispatch office and work dispatch, considering you can't do anything here." They saw that I could handle radios and could keep track of everything. At that time, they had a shortage of people to do my job, and they still do because not many people want to do it. It's stressful. You've got to be quick. Plus, you have to be a qualified division supervisor or an incident commander type three to even get into the class, which is a two-week course they give in Redding. Then you have to have two to three years of actual field experience to complete the training. It's hard to finish because it's hard to pick up assignments. Then you can miss a year, and that's it. You'd have to repeat and start from scratch.*

*The biggest problem I've had is when I've lost track of an aircraft—they're either not answering because of radio problems or they've gone to the wrong frequency. So what runs in the back of your mind is that you have all these aircraft in close proximity, and I don't know where that one's at. You have to figure that out right away because you might be turning into him. Mid-airs are really rare, but they do happen. A simple error, like the wrong frequency, can do it.*

*Then of course, there's the strain on family life. It's a lot easier now because they have limits about how long you can be gone. Up until 1987, you could be gone six or seven weeks. Then they cut it down to three weeks. Now it's two weeks. You can stay an extra week if the situation warrants. It's tough. Especially if you're away for a couple of weeks, then come home for a day or so and then have to go back. There are a lot of marriages that don't make it. It's just not worth it—on either side.*

# 7

# FIRE

## WHEN, NOT IF

*Lightning fires have always burned in our hot and dry summertime landscapes and always will. It is up to the public to determine whether the wildland fire will be gentle and beneficial affairs or raging holocausts that devastate the vegetation, soils, watersheds, and wildlife, and sometime spread so relentlessly that they kill people and destroy homes.*
*—Harold H. Biswell*

S uppression has been the classic approach to fighting fire for one hundred years. This successful, definitive strategy that has squelched fire's advancing over the landscape has consequently allowed fuels to build up to unmanageable levels and, as Pyne points out, "can also be expensive, dangerous, ecologically damaging and self-defeating." Suppression will not go away. It cannot be entirely abandoned as a method for fighting fires. There are, however, things that can be done to reduce the devastating effects of wildfires in addition to cutting costs on suppression, what Pyne refers to as "crafting an environment that yields controllable fires." For the private landowner, that includes, at the very least, brush clearing and tree thinning.

But there is a lot of resistance and inertia among residents to reducing fuels on their properties. And they certainly don't want public officials dictating what they should do on their own land, even though inevitably they will be confronted with fire. "There are a lot of reasons why people don't clear," said the Cal Fire battalion chief at Shaver Lake, Mark Glass.

*We have those who don't trust the government. We approach them, explain the Public Resource code that lets them know what they should be doing. Annually, I'm supposed to go and inspect everyone's property, to make sure they're doing what they should be doing. But they want to enjoy the beauty and live here with the forest right up to their yard and have all this privacy. It's what drew them here.*

*But they really hate it when it gets black. And they cry and they moan and they groan because it's been devastated by a fire. Well, I tell them, if you just mow your weeds down, that would make a big difference. We don't ask you to strip it down to bare dirt. Keep the weeds low to the ground, put some fire resistive plants around your house, clean the pine needles or leaves off the roof, clean out your gutters. You'll enjoy your place a lot longer, naturally, instead of black.*

*We do an annual inspection. We try to get out to everybody. In my area, we have about three thousand structures [up at Shaver] that we need to inspect. For me to get out and do three thousand inspections and train and go to fires and go to medical aids and other emergencies and all the other things we have to do in our daily lives, we're lucky if we get three quarters of them done.*

*Maybe we need to show what happens when you have a fire. Instead of the news reports saying it was devastating, show a few cats or dogs that are laying there all burned up, a few homes laying in cinders or a couple of cars and trucks gutted. But maybe people really don't think it could happen to them.*

A question for Sierra foothill residents: What was your initiation to fire? Was it a robocall from the county telling you to evacuate because the "fire is coming your way?" Or did you wake up one night to a sense of something wrong, only to see a strip of flame creeping over the ridge along the edge of your property? Maybe your introduction happened during a wildfire that choked the air with smoke for two weeks, and firefighters became a fixture in your town.

Whatever your personal experience, if you've lived through a summer in the foothills and mountains adjacent to the Sierra National Forest, you have not evaded fire. And you have not been spared the warnings of foresters, fire ecologists, firefighters and government officials that fires are inevitable. You are aware of the buildup of fuels. You have a sense that wildfires are getting hotter, larger and harder to fight and that firefighters are having an increasingly difficult time keeping fires from spreading into private communities.

Anita and Glen Pickren did not evade wildfire. They knew about the importance of reducing fuels around their home from their fire experience living in Southern California. So when they moved to Mariposa, they signed up for a forestry management program that was available through Cal Fire at the time and were able to clear fifty acres of their land. When the Carsten's fire in 2013 started half a mile from their property, the firefighters not only were prepared to defend their home but also used their property as a staging area. The Pickrens V-shaped pond provided fifty thousand gallons of water an hour to the firefighting campaign. Three helicopters flew in, one behind the other, every forty-five seconds to draw water. "We were entertained with an air show, just sitting here in our house watching the helicopters fly in and out," said Glen. Anita commented on their experience hosting the firefighters. "We just loved them," she said. "They were so strong and brave, just wonderful people." Both Anita and Glen agreed that while the firefighters contributed to the calmness they experienced, it was prayer and faith that helped them achieve a sense of peace during the ten days the fire burned.

"In the old days," said longtime firefighter Smiley Tierney, "fires didn't last more than a couple of days. Now they're getting longer. We got better and better at suppressing fires and the brush started growing, and now we've got these raging fires." Smiley pointed out his window. "When you look out in our backyard, you can see how everything is cleared. But if you look way back there's a wall of brush. Well that wall of brush used to be up here by the house. And if a fire came before we cleared it, well you know what would have happened. It took us ten years to get everything all cleared out."

"More and more, we are involved with private homes as opposed to open lands," said Karen Guillemin, Cal Fire prevention officer. "My dilemma right now," she explained, "is dealing with people who don't want to remove a single tree from their property. People cherish their views of what they consider to be untouched forest. They want it to remain natural. How do I explain to them that because they put their house there, it's no longer natural, that they have upset the balance that Mother Nature provided the world?"

"Part of it is the romantic ideal of people who are retiring to the mountains," said Margaret Schroeder, longtime resident of Midpines, a small community outside of Mariposa. "They have this idea that they are nestled within the forest and they have the trees all around and the wildlife, and that's how they want it to stay." What they don't consider, she fears, is that when there is a fire—"and fire will come," asserted Margaret—their precious trees and their tall red-barked manzanita, their deer brush and all their flowering shrubs will not survive.

If they could picture their land blackened, an eyesore for years to come, imagine their beautiful view damaged and the value of their property diminished, would they change their minds? Perhaps. But for those who give up their city homes and settle into their mountain paradise, that scenario might not be entirely clear at first. It might take some time for those people to grasp their special relationship to the land.

"I think you have to have history to understand," said Margaret's husband, John, a fourth generation landowner. For him, fuel reduction on his land is unquestioned. It's in his blood. He and Margaret live on the property that was settled by his great-grandfather. "My great grandfather came on a stage from Wisconsin in the early 1860s," said John. "Like so many pioneers, he learned to live by his wits." The cabin John's great-grandfather built in 1869 still stands and now holds an eighteenth-century pipe organ, which John is restoring. The Schroeders live in the home built in 1902 by his grandfather. It's all in their history, a family connected to stories of mining and gold and goats and apple trees.

"The people who lived around here were people of the land," John explained.

> *They knew that fire was as natural a part of their environment as was the sun, the snow, and the wind. They cleared their brush and thinned their trees, just as we do today. With the current group of people who live here, the old families have died off, the descendants of those families have moved away, so there's a kind of societal amnesia. They've lost the knowledge of land management that pervaded the community, and they don't understand that it is their responsibility to do it. Plus, they don't know how to do it. It's going to take decades before people will finally get it. I hope that people won't have to experience the worst of the fire's ravages before they do.*

Diann Miller—secretary for the Foundation for Resource Conservation, a nonprofit organization involved in fuel reduction through chipping—also expressed concern for the attitudes of newcomers to her area in North Fork and for their lack of connection to the land.

> *We have people coming here from Los Angeles and San Francisco who love to look out their windows to the distant forest. They're from the blacktop forest. They don't understand that we have to take care of what we have here. They don't understand that their stewardship goes beyond their homes or neighborhoods, but spills over into the trees, water, streams, the whole area.*

Diann, who lives in on the property her family homesteaded 150 years ago, expressed her desire for her children and grandchildren to become part of the community in North Fork. "I want to pass this land down to them," she said. "I want them to become part of the history of this land, to care for it so it's here for their children."

Karen Guillemin believes part of the problem is that people have lost a sense of community, of feeling responsible to others not only to themselves. "Everyone thinks a little bit differently, and luckily it's still the good old U.S.A. and we all have our own opinions," she said.

> *But when your opinions affect others, it could affect others in destruction of their property, or worse, in a fatality. That's where they have to step up and take responsibility for moving into the wild land. [The year] 2013 was a terrible year, terribly deadly for firefighters. Fires were huge that year. The life loss has been intolerable. The public feels bad, and I'm hoping some good can come out of it. Maybe they'll begin to understand their role in helping prevent future tragedies. Eliminate the brush. It's not natural. Be responsible. Think.*

Perhaps dollars-and-cents arguments might get people to sit up and take notice, suggests Karen. She tells of a landowner who was welding on a one-hundred-degree day with no proper clearance.

> *Some brush caught fire and burned his property. It could have cost him a bundle. I could have issued him a citation. Or air quality control could have come and fined him. There were old tires and a lot of equipment and old cars that were burning, very toxic. The fire started at 10:30 and required attention until 8:00 that night. The suppression required two tankers, two helicopters and hand crews. He could have been fined for the cost of recovery, the entire suppression effort. It could have been a catastrophic fire situation. Luckily, it didn't spread to his neighbors. Damage to someone else's property when you've started a fire carries its own liability. So, in addition to the costs of damage to his own property, a person could be hurt in his pocketbook with all the other areas of liability.*

While there is a lot of energy around educating the public on fire prevention and building community, whether people will take in the information and act on it seems to be dependent on a number of things besides the actual danger involved. For one, sometimes, local communities end up with an adversarial relationship to public agencies, particularly the Forest Service, and they don't want to listen to its messages.

Even when people don't take an adversarial stance or are not critical of the Forest Service, they might not be receptive to the information they receive. A study done recently through the Pacific Southwest Research Station looking at the way people receive, interpret and reconstruct wildfire educational programs found that individual homeowners' perceptions of risk often deviated from the risks conveyed by agency educational programs. While people accepted the threat of wildfire in their area and they knew that fires might burn their homes and significant areas nearby, many did not choose to act on that threat. The majority had externalized the threat of wildfire—it's not about me—and many felt little responsibility for their role in minimizing or suppressing fires.

To help people internalize the fire threat—making it more personal and relevant, thereby increasing the chances that the homeowner will take responsibility for alleviating the threat of fire—the researchers recommended that educational programs about wildfire should be stated in terms that are relevant to homeowners' lives. Is it possible that the usual dramatic wildfire photos with flames shooting into the air, out of control, only reinforce the externalization of the experience, serve to distance the viewer from the experience? Perhaps fear is not a sufficient motivator to action.

John Schroeder is a strong advocate of community projects that can motivate people to work together to reduce fire hazards on their land. "There were always cooperative efforts among the people that lived in this area," he said, "and there's no reason we can't bring that community spirit back." As an associate member of the Mariposa County Resource Conservation District board of directors, John was a major player in a fire-ready workshop that took place in Mariposa in October 2014. The idea of the program was to engage people in assessing their properties for fuel hazards and getting them involved in calculating the different levels of hazards on their own properties and within their community. By doing hazard assessments based on such criteria as slope, topography, vegetation and other physical characteristics, the community would then set priorities for fuel reduction projects.

Education about fuels and fuel reduction is one of the main jobs of the Foundation for Resource Conservation. Diann Miller's husband, Jere, has worked many years for the Sierra National Forest as a fuels technician, measuring fuel loading in the forest and calculating how hazardous it is. "I would go out and take samples of key fuels that have high ignition possibilities, weigh it, figure out how much moisture it contains, what it would take to ignite it," said Jere. The work their organization does helps reduce the hazard of fire, assisting people in understanding the difference between ground fuels and ladder fuels and combustibles that direct the fire into the tree crowns or into a house.

"Sometimes, a ladder fuel can be your house. Your deck can be a ladder fuel," said Diann. "If you have fire running up a hillside and it starts rolling underneath that deck, begins to preheat it, it can actually explode. You can't do anything about your deck, but you can reduce the fuel loading down the hillside. You can put different, less flammable plants. You could have a fire resistant barrier in place in front of the deck. But remember, California law requires one hundred feet of clearance around your home—thirty feet immediately around your house and a reduced zone of seventy feet."

What is defended in a defensible space? Two things; your home and the safety of the firefighters. It may come as a surprise to some people that firefighters are not expected to put themselves at risk trying to save your home. In episode seven of Harold Hendrix's *Dragon on the Land*, a series of vignettes describing what people might experience during wildfires, Cal Fire Shaver Lake battalion chief Mark Glass lays out several scenarios describing which homes might be defended and which would not.

> *When crews arrive in their tankers and there's fire all around and people have been evacuated, the job of the fire boss is to decide which homes they are going to defend and which they are going to leave alone. A big part of the decision-making process will center on the amount and quality of defensible space. In the story, Glass refers to this process as "triage." If they come across a neighborhood and structures with decent defensible space clearance, even if they are somewhat directly in the path of the fire, those structures would be considered savable, and the fire boss would send crews in to defend that home. But, if there is a lack of defensible space and they are in the line of fire, there will not be an attempt to go in.*

Karen Guillemin describes a real-life tale of a triage situation. "The John West area of Oakhurst is a concern because when I drive down there I'm just looking at the fuels and I try to imagine what it would be like if there were a fire and I had to send my crews down there," she said.

> *It would be like driving through a tunnel of flames. Are you going to send a fire engine with a crew into a tunnel of flames? Do you want your family to try to escape through a tunnel of flames? So the public that chooses—and it's your choice in America—to live there, then you need to get together and you need to make sure that you and your neighbors have taken responsibility for creating the defensible space that will allow the fire trucks to enter safely to fight fires.*

"This is a new understanding of the reality of fighting wildfires," explains Richard Bagley.

*Looking forward down the road, it might take fifty years, but people will eventually get it. If you're going to live in a fire-prone area, you need to be self-sustaining. We don't ask the government to come throw a tarp over our house when it's going to rain. It's a natural event and we accept that we need to put a roof on our house. So you need to build your house so it will withstand fire, and you have to maintain the landscape around it.*

## ON ANOTHER NOTE

## *The Firefighter's Wife: One Woman's Story*

*Being a firefighter's wife is not easy. I sometimes felt that's all I was. Their job becomes so important and sometimes you give up part of yourself.*
*—Lori Oliver Tierney*

Mountain people appreciate their firefighters, especially when there's a wildfire threatening their town. Signs expressing love and gratitude pop up all around. People give firefighters the thumbs-up as they drive through town, approach them as they stand in line at the bank or are having coffee at a local café just to say thank you.

People do all kinds of things. "All the time we get that," said Mark Glass.

*I appreciate the signs. Signs along the road are neat. A plate of cookies now and then is great. It's fun and I appreciate that. But I've had situations where five engines and other staff—maybe sixteen people—are sitting in a restaurant eating and the manager comes over and says, "Your bill's been paid." That wasn't a cheap meal. That's when I get to feel a little uncomfortable. I try to be a humble person, but I signed up to be a firefighter. I knew that what I was doing was helping my community. A simple thank you, a smile, a thumbs-up, that'll do it. People shouldn't have to spend a lot of money to say thank you. People pay their taxes. They've already bought my meal. If they want to spend their money, they should donate it to the Red Cross or a food bank that takes care of people.*

What people may not realize about those firefighters they're bestowing gifts on is that often when the job in their town is done they will turn around and head to another fire somewhere. It's not uncommon practice for this to happen, particularly in a drought year like 2014, when wildfires burned continually in California over the course of the summer.

But the families of the firefighters know this. It's what they've come to expect.

"It's not easy," said Lori Oliver Tierney, who has been married to her firefighter husband, Smiley, for thirty-five years.

*When people say, "Wow you've been married thirty-five years?" I laughingly tell them, "Yes, but I've probably only lived with him about twenty-five of those years."*

*Many months Smiley was not around. Sometimes he'd be gone two or three months at a time. Some of that is good, actually. You get to do things you want to do, aside from being with your husband. You really don't want to be the wife of a firefighter if you do not have a sense of yourself. They're not around a lot of the time to feed that, so the woman a firefighter marries has to be a woman in her own right, not a girl.*

*There have been many, many divorces in the Forest Service. It used to be that a firefighter could sign on to a fire and continue to sign on as long as they wanted. Now after two or three weeks, they have to take a break. And that was a good thing. People were in trouble being separated for such long periods of time.*

*There were times when Smiley and I would joke with each other and say, "We'd better not go to that function for awhile. Five couples are gone. Maybe it's in the water." You have to have a sense of humor.*

Lori got her degree in parks and recreation and was working as a ranger for the Sierra National Forest at Big Sandy campground. At that time, she was living at Bass Lake, and that's where she met Smiley.

*He was the weirdest man I ever met. This guy with long hair, a climber, a crazy man. He drove motorcycles. I thought, my gosh, I'm not interested in a relationship. I want to be free. I was twenty-five. I wanted to travel.*

*But, to be honest, we fell crazy in love. And back then, living in a small town, Smiley didn't think it was a good idea for us to just live together. Today things are so different.*

*So, we met in May and got married in October, and I've never regretted it. Of course, there were moments when I did, but not over the long haul.*

After they married, Lori stayed with the Forest Service doing various jobs: fire prevention tech, back country ranger, timber marker. Then they began to talk about having children, and finances became a serious issue.

*In firefighting, it's either feast or famine. The Forest Service firefighters are the lowest paid of all the public agencies. They do make a lot of overtime when they're out on a fire, so that helps pad the bank account during those lean winter months when they might be laid off and have to live on unemployment. During those times, Smiley would go anywhere he could get work. He'd cut down trees for fifteen dollars an hour if he had to.*

*So I decided I needed to get a job where you could count on the salary, where you knew what the paycheck was going to be. With his you couldn't.*

*I came to a decision that, if we were going to have kids, I needed a job that was around home. So I gave up my dream of being in this special job. And what did I choose? Teaching. It wasn't my plan to become a teacher. It's so funny. I had this idea in my mind that I'm such an independent woman. I didn't want to get married. I wanted to travel and have this special kind of job working in the wilderness. And then I got married and went into teaching.*

*As it turns out, I think my actual calling is in teaching. I think I fought it because it's such a woman's job, nothing sexy about it. But I love it. And I didn't have to give up my plans to travel. Smiley's a really good teacher. So the Forest Service used to send him all over, and I would go along. He taught in Greece and Italy. He even got to teach in Africa. He was always involved in his work, so I got to go around by myself and visit all these wonderful places. I've had a wonderful time traveling.*

Staying home with the children when they were young was difficult. Lori admitted that she got lonely a lot. But Lori's issues with Smiley went beyond his firefighter life and seemed to mirror some of the normal day-to-day spousal clashes.

*It's a partnership, you know. Not only does the wife have to be understanding, the husband has to also. But, figuring out how to be helpful at home wasn't his thing. The understanding touchy-feely stuff, that's not Smiley. He's a firefighter type through and through, craves excitement, activity. He's good at cut and dried things, like making life and death decisions.*

*He told me he couldn't read my mind. I had to talk it through with him. He asked me to write things down, to tell him what I needed. So I would*

Lori Oliver Tierney with her fourth grade students. *Courtesy of Lori and Smiley Tierney.*

Smiley with students in Africa. *Courtesy of Lori and Smiley Tierney.*

*write letters and tell him all the stuff I do when he's gone and what I needed from him when he's at home.*

*One winter when he was around, I came home from school one day and told him I didn't want to do homework with our kids. I'm a teacher, and I do enough of that. To his credit he said, "I'll do the homework all winter long. You can be the mom and you don't have to be a teacher." Thank you very much, Smiley. I did not do homework at all that season.*

*When our oldest boy turned ten, I told Smiley I wanted him to take a vacation with us during the summer, which meant he'd have to sign off on some fires. We went a lot of places with the boys, and we had great fun.*

*That's one thing. We always could have fun together. We were friends.*

# On Another Note

## Smokey Bear: Zip Code 20252

*People have heard Smokey the Bear's fire prevention message for years and years, but hardly anything on the details of using fire constructively or about the idea of working in harmony with nature.*
—*Harold H. Biswell*

*Only you can prevent wildfires.*
—*Smokey Bear*

Smokey Bear is famous, recognized worldwide as the icon of fire safety. His fans send him so much mail that, in 1964, the United States Postal Service assigned him his own zip code. Only one person in America has that honor, the president.

But Smokey is also infamous. In certain circles, he has become a symbol of poor forest management and is linked to the public's fear of fire and lack of understanding of fire's natural usefulness. Smokey is so much a part of the consciousness of the fire community that he has made his way into scholarly books written by fire ecologists, researchers and historians.

Smokey's story is told in black and white, a right and wrong tale of what his critics refer to as "one of the most successful public relations campaigns in U.S. history." But these days, such a polarized message can't work. The public is being called upon to become participants in planning and creating defensible space around their homes. They are being asked to give up their fear and learn about fire behavior and to apply their knowledge in creating fire safe communities. What is needed is a message that educates the public to respect the power of fire to destroy but also to understand and welcome its ecological powers.

The public's education about fire as the enemy began long before Smokey came on the scene in 1944. Posters from 1911 show fire as death riding through the forest. "Fire the destroyer: Keep him out of the California woods," one poster reads. A 1939 editorial in *American Forest*, an edition devoted entirely to fire prevention, cautioned that Americans needed the will to fight fire. "This issue of American Forest," the article states, "is dedicated to the creation of such a will," pairing fire prevention with courage, religious faith and patriotism. Specific connections are made to early patriots of America, encouraging the public to become the "Paul Reveres on the public opinion highways of our country" in spreading the word about fire prevention.

Early in World War II, fire the enemy became enmeshed with America's real enemies when Japan began using fire as an attack strategy along the coast of California. In 1942, Japan sent nine thousand firebomb balloons over the West Coast. Several hundred fell over twenty-six states and provinces from Alaska to Mexico. Five children and a teacher were killed after one of the firebombs detonated. It was then that the Wartime Advertising Council began producing posters with lurid images and messages geared to demonize the Japanese and the Nazis, coupling fire prevention with national defense. "Tokyo loves an American Forest Fire" is the caption of one such poster. Other messages were:

- Fires Fight for the Axis [decorated with a Nazi swastika]
- Our Carelessness, Their Secret Weapon: Prevent Forest Fires
- To Speed Victory, Prevent Fires: Keep the Home Front Strong

So by the end of World War II, the connection between fire prevention and public responsibility was set. In 1944, Smokey appeared, and while the posters lost their ghastly appearance, the basic message they carried was the same: fires are destructive and need to be prevented. Over the decades, Smokey's messages have changed, but all have similar reminders and suggestions.

- Prevent wildfires
- Help people be more careful
- Protect our forest friends
- Be extra careful during certain years [like during a drought]
- Care will prevent nine out of ten fires
- Shameful waste weakens America

Some of Smokey's messages overstep the "be careful" idea. An example: "One match, that's all it takes to destroy a forest and every creature great and small that lives there." This is more than an exaggeration. It's simply untrue. But a fearful public is vulnerable to believing such a story.

Smokey, however, is only a messenger. He doesn't care what he says. He will deliver whatever spiel he is fed. Can Smokey's message change? If the general public is brainwashed by Smokey's message, as has been suggested, could the same marketing techniques be used in a reverse brainwashing campaign?

"We have some people who think that fire is bad and that's not true," said fire prevention officer Francis Adams.

> *Fire that kills or damages property, that's different. But when we talk about the forest, if Smokey were standing on the billboard and planting new trees, using him as a campaign for good replacement of what resulted from a dangerous fire situation, his message would be, "OK, now the fire danger is over, we will plant new trees. Everything would be clearer and cleaner because the ecosystem is renewed." We could do an ecology campaign with Smokey.*

That idea may not be so far-fetched. In 2004, when Smokey turned sixty, the comic strip Mark Trail, a syndicated cartoon that began in 1946 and focused on environmental issues, included a message from Smokey that said, "I don't promote the suppression of wildfires or prescribed fires...my message is to prevent CARELESS wildfires! Prescribed fires can be beneficial to plants and animals and to prevent wildfires if they are done under SUPERVISED conditions." So, possibly, as the public becomes more familiar and comfortable with fire ecology and prescribed burning, it might begin to accept complementary messages from Smokey. It's fun to fantasize Smokey up on a billboard under the caption "Burning Can Prevent Wildfire."

# A Personal Story

## Evacuation: Time to Go

The call came at five o'clock. "You have been evacuated. Leave your house. The fire is coming in your direction." I'm prepared. My fire "go box" and computer are already in my car. I throw some clothes into a suitcase. I scuffle with my two outraged cats and finally get them into their carrying cases and then, per usual, run madly around searching for my purse and car keys. Finally, everything's packed, and I'm off to the evacuation shelter.

By nine o'clock that night, the Red Cross has registered us, fed us and set up cots for thirty overnighters. My cats are ensconced in an air conditioned "kennel" set up in a room apart from the evacuees by the Central California Animal Disaster Team (CCADT), an organization formed in 2011 that works side by side with the Red Cross during disaster relief. My cats, along with twenty other unwilling cats and dogs, are making a ruckus, unaware of the fire danger all around and unmoved by the luxury of their five-star accommodations. The staff of the CCADT, patient, resourceful, skillful, somehow manages to calm the refugees—the four-legged and human type—and we are all settled into what will be our routine for the next three days.

Then, to my surprise, I look around and wonder, "What do I do now?" I am bored. Bored? Who could have imagined that amid all the panic and confusion, not knowing whether I would have a house to go back to, I would experience boredom? I blame it on the Red Cross and CCADT. Too efficient.

The next morning after a briefing on the status of the fire, a breakfast is served. The choices include eggs, bacon, pancakes, cereals, juices, fruit, coffee, tea and cocoa. I make another visit to my cats. Then, again, there is a sense of what do I do now? Sweep and mop the floors. Wipe down some tables. Talk to some folks. Bring some coffee to a fellow evacuee. Try to calm a worried woman. Another hour gone by. Another visit to the kennel. Everything under control there. Nothing to do.

A moment of panic. At last! I think about my house. What's happening to it? Pictures in my mind of it sitting alone, on fire, the feral cats I feed scurrying about, looking for food. They say that feral animals know how to escape fire. I know they will be all right.

I drive back into town. The roadblocks are still up. I can't get through. I try to think of a way around them, but the road to my house is pretty well closed off. Back to the shelter. I learn from the notices tacked on the wall that they've named my fire the Junction Fire. A Red Cross representative is interviewing a couple. "They've lost their home," one of my new evacuee acquaintances whispers. She didn't have to say a word. The expression on the couple's faces, the tears in the woman's eyes, they tell the whole story.

I call my home, relieved to hear my voice mail click in and then the beep. I leave a message. "You're still there," I say. "I'm thinking about you. I miss you and hope I'll see you soon." I would call my home and leave love messages three more times before I was allowed to return.

Evacuation is doing crazy things like talking to your home. Evacuation is being suspended in a limbo state, not here, not there. There's nothing you can do, what with all the accoutrements of normal life sequestered twenty miles from where you are. The help and company of friends, a meal, a visit, an overnight stay—these all help keep you rooted, earthbound. But it's always with you: the house, the fire, the not knowing. You never fully engage. In your mind, you are sifting through the ashes of your home, searching for things you had to leave behind. You are planning. There's the insurance problem, temporary living, rebuilding. Maybe it's a good thing. Maybe you're preparing yourself for the worse. You're safe, you tell yourself. The rest is in the detail.

Evacuation is also about miracles, some small, some large. In the shelter, people related their stories of miraculous things happening during the fire. We heard about the propane company's office building burning up into a cinder, yet the giant propane tanks just ten feet away remaining untouched. "If those tanks blew," one evacuee said, "there'd be a crater four miles wide in this town." (We would learn later, after everything had settled down, that the miracle was actually the Cal Fire crew's skill, ingenuity and courage that prevented the tanks from blowing up.)

I experienced some miracles of my own during those days. I almost had to leave one of my cats when she scratched her way out of my grip as I was putting her into her carrier. She ran and hid under a couch. I couldn't coax her out. Not even her favorite treat could grab her attention. I moved the couch, and she moved with it. I begged her to come out. In desperation, I walked out of the room thinking I would have to leave her. Then back into the room and one last try. In my most authoritative voice I called out, "You've got to come out, now!" And she did. Like an obedient puppy dog, she walked right up to me. But she's a cat, and you know what they say about herding cats.

Then there was my phone charger miracle. My children kept calling. They were worried. Should they come? Is there anything they could do? What's going on? I had to rush them off the phone because my cellphone charge was running down. How could I have forgotten my cellphone charger, I berated myself. Then the solution came to me. I drove to the business center in town and asked if they could charge my phone. The sales person took my phone to the back and emerged with a charger. "We have tons of these back there," she said, handing me one, no questions asked. We chatted for a while, and I found out she too had been evacuated. We were kinsmen.

This act of kindness was one part of the miracles of these days of evacuation, the giving, generous community in which we live. Local businesses asked, "What do you need?" They opened up their doors, provided services; local citizens and organizations showered us with food, clothing and personal items—everything from toothpaste to Huggies. As one volunteer said, "We told people we needed deodorant, and fifty deodorants showed up."

And then there were donations of hours and days of personal time, people arriving and asking, "What can I do? Do you need anything?" These people—quietly, efficiently and in good humor—going about their work. They take time from their own lives, from their jobs. I think of a couple, Kelly and Dan, with nonstop kitchen duty, cooking and serving meals, cleaning floors and tables, ten, twelve hours a day. What makes people do that? "I was just raised with the idea that when you serve your fellow man," said Kelly, "you're serving God. My dad taught me that. My mom taught me that. I taught it to my children." Amen.

Over the three days of evacuation, I learned that there were evacuees of all kinds, as varied as any gathering of people. But the ones that captured my imagination were those who, when told to evacuate, dug in their heels and stayed, determined to wait it out and protect their homes. Or there were those who found their way back to their houses, traveling as much as fifty miles out of their way over forest roads, to retrieve pets left behind. Why do they do that? Does it work? I try to picture myself on the roof of my house, hose in hand. Where would I start? What would happen if the water pressure ran low?

"We come across those kinds of evacuation issues with every fire," said Karen Guillemin, Cal Fire prevention officer. "Some people won't leave their homes, and we have to force them out of there. They don't think about all the equipment coming in, huge fire engines, bulldozers, water tenders. We need a lot of area to work. Helicopters come in and drop buckets of water. You could be knocked off your roof. You need to be out of there." Karen admits that, under some circumstances, people can save their homes by hosing their roofs, but she urges people to educate themselves about the right conditions to do it and then to know when it's time to leave.

After I am finally allowed to return to my home and find everything as I left it, I cannot settle down. It's like having an engine running inside me. I drive around town, walk along charred hills. I see places where the fire has burned up to the hill behind a house and stops its advance. I see a property with the skeleton of a shed twenty feet from a house that stands untouched. How does a fire decide where it is going to go? I visit the site of the propane company. The office building lies in charred ruins. The enormous white gas tanks sit quietly nearby. I am overcome with gratitude.

## PART IV

# HISTORY IN THE MAKING

In 2009, Congress passed the Omnibus Public Land Management Act, which includes as one of its provisions the Collaborative Forest Landscape Restoration Program, a science-based, landscape-scale project aimed at reducing the risk of wildfire through ecological restoration.

In the Sierra National Forest, the Dinkey Forest Landscape Restoration Project (DFLRP), made up of twenty diverse stake-holders, was chosen as one of ten collaboratives nationwide—three in California—to design, carry out and monitor a ten-year ecological restoration program. The Dinkey project consists of 154,000 acres on the High Sierra Ranger District.

At this time—2014—six years remains for them to accomplish their goals.

# 8

# COLLABORATE, COLLABORATE

## *THE DINKEY LANDSCAPE RESTORATION PROJECT*

Congress has included ecological restoration and science-based, landscape-scale planning in its prescription for restoring the health of the national forests and reducing the risk of calamitous wildfire. But it went a step further when it added collaboration as part of the treatment. Why did it do that?

It appears that Congress has sent a message to the American people that the national forests are not here for only our pleasure and exploitation. Rather, we are being asked to learn about forest health and to step up and get active in the process of restoration. By insisting on collaboration, it seems that Congress has also sent a message to the Forest Service. Building a productive relationship with the American people is as important for the future of the national forests as any scientific or ecological planning.

Craig Thomas, conservation director for Sierra Forest Legacy and one of the designers of the original Dinkey plan, believes the collaborative process is necessary. "We've got to have more social unity in terms of how we manage our natural resources," he said. "And I work on this every day. People need to come together and face the challenges and fears of dealing with people who don't agree with them. Because I frankly believe we're toast if we don't."

Four years ago, a group of stakeholders took up the challenge and formed the Dinkey Collaborative. "We have a broad range of individuals and organizations represented in the collaborative," said Dorian Fougeres, Dinkey's mediator. "We have the forest products industry, Southern California

Edison, tribal involvement, the environmental community, groups involved in recreation, private landowners, business owners, academia, nongovernment organizations, private homeowners. There is staff from the Sierra Forest as well and from the Forest Service's Pacific Southwest Research Station."

Dorian is the director of the Southern California Office of the Center for Collaborative Policy, a unit of the College of Social Sciences and Interdisciplinary Studies at California State University. The mission of the center is to build the capacity of public agencies, stakeholder groups and the public to use collaborative strategies to improve policy outcomes. Dorian has worked with regional and statewide collaboratives in California.

"This is the best group of stakeholders I've ever worked with," he said. "My impression of the group is that they operate with a high level of capacity in terms of getting work done. They also stand out as a regional project for the high level of complexity in the committee and sub-committee work they do. The group is technically oriented on a level of detail among the highest I've seen. And the dedication of the Forest Service people and their commitment to the project is very high."

"The thing I praise in Dinkey," Craig said, "is the respect that has been built and the barriers that have been broken down. That's a social goal for me personally, and I think it's the way the environmental community needs to do its work. Get into these collaborative settings, bring our knowledge and skills, and hear what other people have to say. We'll never sacrifice what matters to us, but I think there's a whole big bunch of room to work in there."

One of the areas where Dorian believes the group has made great strides is in the socioeconomic indicators, "things like public perception of the Forest Service, public value of forest management, how to handle campgrounds, how Dinkey meshes with recreational interests," he said. "As far as tribal involvement in the collaborative, we are resolving a variety of issues related to tribal needs, and their involvement has been consistent and ongoing."

Another area where the collaborative process seems to have had a positive impact is in the amount of litigation and appeals of projects planned by the Forest Service. Carolyn Ballard is the fire management officer at the High Sierra Ranger District of the Sierra National Forest. She has noticed a significant reduction in the amount of appeals lately and attributes this to the Forest Service's policy of keeping the public informed and involved.

*The process goes something like this: in the Forest Service, once we come up with a plan and write an environmental document, it goes out to the public for review in draft form. We get letters and letters and pages and pages:*

*what about this science, and did you look at that analysis, or we don't think you analyzed that well enough. We make changes and go back with the final document.*

*Then if the forest ranger or supervisor makes a decision about moving ahead, the public has an opportunity to appeal the project to the regional office. At that point, the project is reviewed to see if we addressed the public's concerns. Did we answer their questions? Did we do what they asked? Was what they asked within the scope of the project? Did we do what we said we were going to do and keep the public in the loop? That's when they can approve or rescind the decision. If they decide to move ahead, and the public still doesn't like it, that's when they can go to the courts. We've been very successful lately without having to go to the courts to have them appealed.*

For the Dinkey project, Carolyn's office has full implementation responsibility to conduct the on-the-ground work once the planning and environmental processes are completed.

*The way the Dinkey collaborative works, all interested parties are involved in bringing forward their thoughts and concerns and desires upfront to design projects that everyone can live with. Sometimes the negotiating is really tough. People have strong feelings and opinions about things, whether it's protecting the owl or the fisher, decommissioning a road or recreation use. Everyone in Dinkey realizes you're not going to get exactly what you want. But everyone struggles because in the end we're all looking forward to a healthy forest.*

One of the major challenges facing Dinkey in the future is its being able to come to grips with landscape-scale planning. "It's one thing to plan for 10 acres, or 100 acres," said Dorian, "but when you get to 3,000 acres, 5,000, 10,000 acres or 100,000 acres, that's a different thing. Will the administration in the Sierra Forest be willing to get away from micro-scale projects and go up to landscape-scale planning? We're making progress working with the Forest Service regional office on this, but we have no control over where they are going to apply their funds or what their priorities will be. Dinkey's plans may not get to the top of their list. After all, the district is bigger than Dinkey."

There's a lot of hope and enthusiasm among the Dinkey members, despite the challenges. The collaborative is spurred on by the belief that its work, if successful, could have relevancy for the remainder of the Sierra National Forest and even for other national forests in the region.

For Craig, the question is, with climate change and the drought, with the reality of destructive wildfires, will Dinkey be able to adapt? Will it be flexible enough to adjust to new realities down the line? "If you ask what I fear," he said, "it's failure, that for some reason we won't make it."

# BIBLIOGRAPHY

Agee, James K. Introduction to "The Relation of Forests and Forest Fires," by Gifford Pinchot. *Fire Ecology* 7, no. 3 (2011).

Aldern, Jared D. "North Fork Mono Meadow Restoration, Fire and Water: The Tribe's Land and Water Rights and Tenure." *Indian Land Tenure Foundation: Lessons of Our California Land*. Landlessons.org/Meadows.pdf. 2012.

Anderson, M. Kat. *Taming the Wild: Native American Knowledge and the Management of California's Natural Resources*. Berkeley: University of California Press, 2005.

Arno, Stephen F. *Discovering Sierra Trees*. N.p.: Yosemite and Sequoia Associations, 1973.

Arno, Stephen F., and Steven Allison-Bunnell. *Flames in Our Forest: Disaster or Ruin?* Washington, D.C.: Island Press, 2002.

Barnett, James. "Chapman Called 'Father of Controlled Burning.'" *Faces from the Past: Highlighting Professionals Who Have Made Lasting Contributions in the Forest Industry of Louisiana and the Southern Region*, 2007.

Beaty, Jeanne K. *Lookout Wife*. New York: Random House, 1953.

Biswell, Harold H. *Prescribed Burning in California Wildlands Vegetation Management*. Berkeley: University of California Press, 1989.

Blackburn, Thomas C., and Kat Anderson, eds. *Before the Wilderness: Environmental Management by Native Californians*. Banning, CA: Ballena Press Anthropological Papers, 1993.

Carle, David. *Burning Questions: America's Fight with Nature's Fire*. Westport, CT: Praeger Publishers, 2002.

Cermak, Robert W. *Fire in the Forest: A History of Forest Fire Control on the National Forests in California, 1898–1956*. Albany, CA: U.S. Department of Agriculture, Forest Service Pacific Southwest Region, 2005.

Cochrane, Timothy. "Trial by Fire: Early Forest Rangers' Fire Stories." *Forest and Conservation History* 35 (1991): 16–23.

DuBois, Coert. *National Forest Fire-Protection Plans*. For the U.S. Department of Agriculture National Forests. Washington, D.C.: Government Printing Office, 1911.

Egan, Timothy. *The Big Burn*. New York: Houghton Mifflin Harcourt, 2009.

Goode, Ron. *Cultural Burn*. Tribal Chair North Fork Mono Tribe. Paper Presented to the Dinkey Collaborative, Dinkey Creek, CA, 2014.

Greeley, William B. "Piute Forestry: or, The Fallacy of Light Burning." *Forest History Today* (1920): 33–37.

Gruell, George E. *Fire in Sierra Nevada Forests: A Photographic Interpretation of Ecological Change Since 1849*. Missoula, MT: Mountain Press Publishing Company, 2001.

Headley, Roy. *Fire Suppression, Manual for District 5*. 1916. Reproduction by Jeff P. Prestemon. Albany, CA: United States Forest Service, 2012.

Hendee, Clare W. "Forest Fire Prevention: Progress and Prediction." Presentation at the annual meeting of the Society of American Foresters, Gulfport, Mississippi, 1961.

Hudson, Mark. *Fire Management in the American West*. Boulder: University Press of Colorado, 2011.

Husari, Sue, H. Thomas Nichols, Neil G. Sugihara and Scott L. Stephens. "Fire and Fuel Management." In *Fire in California's Ecosystems*. Edited by Neil.G. Sugihara, Jan.W. van Wagtendonk, Kevin E. Shaffer, Joann Fites-Kaufman and Andrea Thode. Berkeley: University of California Press, 2006, 444–65.

Kilgore, Bruce M. "Fire's Role in a Sequoia Forest." *Journal of Quaternary Research* 3, no. 3 (1973): 496–513.

Krasnow, Kevin. "Managing Novel Forest Ecosystems: Understanding the Past and Present to Build a Resilient Future." PhD diss., University of California–Berkeley, 2012.

Mount, John R. *Torching Conventional Forestry: The Artful Application of Science*. Fresno, CA: Auberry Press, 2010.

North, Malcolm, ed. "Managing Sierra Nevada Forests." U.S. Department of Agriculture, Forest Service General Technical Report PSW-GTR-237. Albany, CA: U.S. Department of Agriculture, Forest Service, Pacific Southwest Research Station, 2012.

Pinchot, Gifford. *Breaking New Ground*. Washington, D.C.: Island Press, 1947.

———. "Primer of Forestry Parts 1 & 2: The Forest & Practical Forestry." U.S. Department of Agriculture, Division of Forestry Bulletin #24. Washington, D.C.: Government Printing Office, 1899, 1903.

———. "The Relation of Forests and Forest Fires." *National Geographic Magazine* 10 (1899): 393–403.

Pyne, Stephen. J. *Fire: A Brief History*. Seattle: University of Washington Press, 2001.

———. "Passing the Torch." *American Scholar* (Spring 2008).

———. *Tending Fire: Coping with America's Wildland Fires*. Washington, D.C.: Island Press, 2004.

———. *Year of the Fires*. New York: Viking Press, 2001.

Rydell, Carol H.L. "Public Face for Science: Starter Leopold and the Leopold Report. Historical Perspective on Science and Management in Yellowstone National Park." *George Write FORUM* 15, no. 4 (1998): 50–63.

Shinn, Charles H. "Work in a National Forest #5: Holding Down a Mountain Fire." *Forestry and Irrigation* (1907).

Shinn, Julia T. "Forgotten Mother of the Sierra: Letters of Julia Tyler Shinn." Introduction and notes by Grace Tompkins Sargent. *California Historical Society Quarterly* 38, no. 3 (1959).

———. "On the Value of a Ranger's Wife." *Sierra Ranger: Quarterly Bulletin at the Headquarters of the Sierra National Forest.* Edited by Paul G. Redington. Northfork, CA, November 1, 1915.

———. "The Ranger's Boss." *Sierra Ranger: Quarterly Bulletin at the Headquarters of the Sierra National Forest.* Edited by Paul G. Redington. Northfork, CA, July 1930.

Show, Stewart B., and Edward I. Kotok. "Fire and the Forest (California Pine Region)." *U.S. Department of Agriculture Circular* 358 (1925).

Stewart, Omer C. *Forgotten Fires: Native Americans and the Transient Wilderness.* Edited by Henry T. Lewis and M. Kat Anderson. Norman: University of Oklahoma Press, 2009.

Sugihara, Neil G., Jan W. van Wagtendonk and Joann Fites-Kaufman. "Fire as an Ecological Process." In *Fire in California's Ecosystems.* Edited by Neil.G. Sugihara, Jan.W. van Wagtendonk, Kevin E. Shaffer, Joann Fites-Kaufman and Andrea Thode. Berkeley: University of California Press, 2006, 58–74.

United States Department of Agriculture, Forest Service. Historical file A, no. A-5. The personal narrative of Roy Boothe, forest supervisor. Washington, D.C.: Government Printing Office, April 1968.

van Wagtendonk, Jan W. Dr. Biswell's Influence on the Development of Prescribed Burning in California. The Biswell Symposium: Fire Issues and Solutions in Urban Interface and Wildland Ecosystems. General Technical Report PSW-GTR-158. U.S. Department of Agriculture, Forest Service, 1995.

———. "Fire as a Physical Process." In *Fire in California's Ecosystems.* Edited by Neil.G. Sugihara, Jan.W. van Wagtendonk, Kevin E. Shaffer, Joann Fites-Kaufman and Andrea Thode. Berkeley: University of California Press, 2006, 38–57.

———. "The History and Evolution of Wildland Fire Use." *Fire Ecology* 3, no. 2 (2007): 3–17.

Wheeler, M. "Fire Storms." *Discover* (May 1994).

Williams, G.W. "References on the American Indian Use of Fire in Ecosystems." Washington, D.C.: U.S. Department of Agriculture, Forest Service, 2001.

# INDEX

# ABOUT THE AUTHOR

Several years ago while on a hike in the Sierra National Forest, Marcia Penner Freedman sat with her friends, resting beside a small lake. When a helicopter flew in, sucked up water into a giant tank and flew out, the conversation turned to wildfire. As the helicopter repeated its routine, Freedman's interest in firefighting was aroused. After that, she began learning all she could about the subject. At one point, she attended a prescribed-fire field trip put on by Southern California Edison Forestry. That's when her interest in learning about wildfire turned to a determination to write about it.

Photo by Gay Abarbanell.

Freedman was born and raised in the New York City Metropolitan Area. Prior to her 1999 move to the small town of Oakhurst outside Yosemite National Park, she had lived for twenty years in Los Angeles, where, in 1995, she received a PhD in educational psychology from the University of Southern California. For fifteen years, Freedman split her career between writing and teaching psychology and child development at a community college. Since her retirement from her community college position, she has devoted herself exclusively to her writing. As a member of the board of directors of the Coarsegold Resource Conservation District, Freedman hopes to use her writing skills to educate the general public about the importance of reducing the fuels on their properties.